Topics in Mining, Metallurgy and Materials Engineering

“Topics in Mining, Metallurgy and Materials Engineering” welcomes manuscripts in these three main focus areas: Extractive Metallurgy/Mineral Technology; Manufacturing Processes, and Materials Science and Technology. Manuscripts should present scientific solutions for technological problems. The three focus areas have a vertically lined multidisciplinarity, starting from mineral assets, their extraction and processing, their transformation into materials useful for the society, and their interaction with the environment.

**** Indexed by Scopus (2020) ****

Carlos Pérez Bergmann

Techniques for Analysis of Mineral Raw Materials

Springer

Carlos Pérez Bergmann
School of Engineering
Federal University of Rio Grande do Sul
(UFRGS)
Porto Alegre, Rio Grande do Sul, Brazil

ISSN 2364-3293 ISSN 2364-3307 (electronic)
Topics in Mining, Metallurgy and Materials Engineering
ISBN 978-3-031-58984-3 ISBN 978-3-031-58985-0 (eBook)
https://doi.org/10.1007/978-3-031-58985-0

This Springer imprint is published by the registered company Springer Nature Switzerland AG
The registered company address is: Gewerbestrasse 11, 6330 Cham, Switzerland

Description of the Handbook

Objective

This handbook presents the content of the course techniques for analysis of mineral raw materials, which aims to present basic concepts in the analysis of raw material characteristics that are defining of technological potential in their industrial application.

The approach to these analysis techniques will be generic and descriptive discussing their main theoretical foundations. Aspects of experimental measurements will be presented, with the aim of providing a contextual view of the techniques, and how to use them as a tool for analysis of mineral raw materials. The description of the techniques will be primarily phenomenological, at the expense of the mathematical formalistic treatment of the physical foundations, necessary for a more rigorous theoretical approach. However, some formulations that support the argumentation of the physical principles and the potentialities of the techniques are inevitable.

At the end of the course, students should have solid knowledge in the main concepts involving the characterization of raw materials and ease in articulating the relationship between the analysis technique and the obtaining of information of industrial and commercial interest.

Contents

Chapter 1
Introduction

Mineralogy defines mineral as a "solid, inorganic and crystalline substance, with a well-defined chemical composition and physical properties, resulting from a terrestrial or extraterrestrial geological process, without human intervention".

Mineral raw materials to be used in the industry in general can be characterized in terms of aspects such as:

(i) Chemical composition: provides the content of each chemical element present in the raw material;
(ii) Mineralogical composition: describes which minerals constitute the raw material (can be qualitative or quantitative);
(iii) Granulometric composition: provides the percentage distribution, in weight, of the particle sizes that constitute the raw material.
(iv) Morphological: informs about the shape and surface area of the particles that constitute the mineral raw material;
(v) Effect of temperature (thermal analysis): informs about the behavior of the raw material (reactions/transformations) when subjected to heating/cooling;
(vi) Microscopic Analysis: assists in obtaining information complementary to those previously mentioned.

Knowledge of these characteristics is essential for the control of the extraction process, beneficiation and industrial processing and will decisively influence the financial results of the companies involved in each of the segments that deal with industrial raw materials.

The characterization of raw materials is, therefore, a fundamental step for the exploitation of a mineral resource, as it provides subsidies to the sizing of the beneficiation process or, even, allows identifying inefficiencies and losses in existing processes.

Numerous are the techniques used in the characterization of raw materials. In the past, the chemical analyses of raw materials were done by wet route (acid attack, titrations, gravimetry, borax pearl fusion, flame tests, among others). In addition to

C. P. Bergmann, *Techniques for Analysis of Mineral Raw Materials*, Topics in Mining, Metallurgy and Materials Engineering, https://doi.org/10.1007/978-3-031-58985-0_1

being laborious, the wet route analysis provided a simple list of contents or merely qualitative results.

With the development of instrumental analytical techniques, such as elemental analysis by X-ray fluorescence (XRF), X-ray diffraction (XRD) and scanning electron microscopy (SEM), the work to obtain data was greatly reduced, but the theoretical complexity of interpretation increased greatly.

The use of techniques for raw material analysis goes beyond manipulating samples and equipment, usually performed by trained technicians. Equally or more important is knowing how to interpret the data obtained in order to make decisions from them. Interpreting is generally much more difficult than using the equipment to obtain data.

Figure 1.1 presents a flowchart of the characterization of mineral raw materials for industrial use. The methodology to be adopted in the characterization of a mineral raw material varies greatly with the mineralogy itself and the inherent properties of the ore, as well as with the objectives and scope of the characterization.

Generally, the characterization of a sample begins with its fractionation, done by density or magnetic susceptibility, for example. Afterwards, phase identification takes place, where specific techniques are used, such as X-ray diffraction, in addition to optical and scanning electron microscopies.

The next step, the quantification of minerals, uses various techniques. Depending on the objectives of the characterization, it is possible to group the minerals according to their response to beneficiation or product specification (for example, iron minerals, including hematite, magnetite, goethite and limonites). One of the objectives of characterization may be to determine the liberation grain size of the mineral of interest (or minerals of interest) in relation to the gangue.

From a methodological-experimental point of view in the laboratory, the selection of physical analysis techniques, the establishment of measurement conditions and the interpretation of results depend, obviously, on the purpose of the study, the information sought, the properties of the minerals. They also invariably require clear knowledge of the origin of the sample (type of soil, genetic factors and processes, collection data) and its preparation (granulometric fractionation, magnetic separation, selective chemical treatments, chemical composition).

Generally, ores are characterized as (i) mixtures of phases with varying degrees of crystallinity; (ii) often, strong cementation and mechanically consolidated grains, quite resistant to disaggregation by conventional laboratory methods; (iii) very varied chemical composition, on a short spatial scale; (iv) wide distribution of sizes and shapes of grains (v) significant proportions of small particles, such as the clay fraction (average diameter, $\varphi < 2\ \mu m$), and (vi) isomorphic substitution of iron in iron oxides by Al^{+3}, Ti^{+4}, Mg^{+2} or Cr^{+3}, to name a few more commonly found cations.

Among the most used analytical techniques in the technological characterization of mineral raw materials, we can mention: chemical analysis by X-ray fluorescence spectrometry and by atomic absorption spectrometry; granulometric analysis and thermal analysis, infrared absorption spectroscopy and microstructural analysis by optical microscopy (petrographic analysis) and scanning electron microscopy, complemented by point chemical analysis by energy dispersive detector. These techniques (highlighted in red in Fig. 1.1) are the focus of this course.

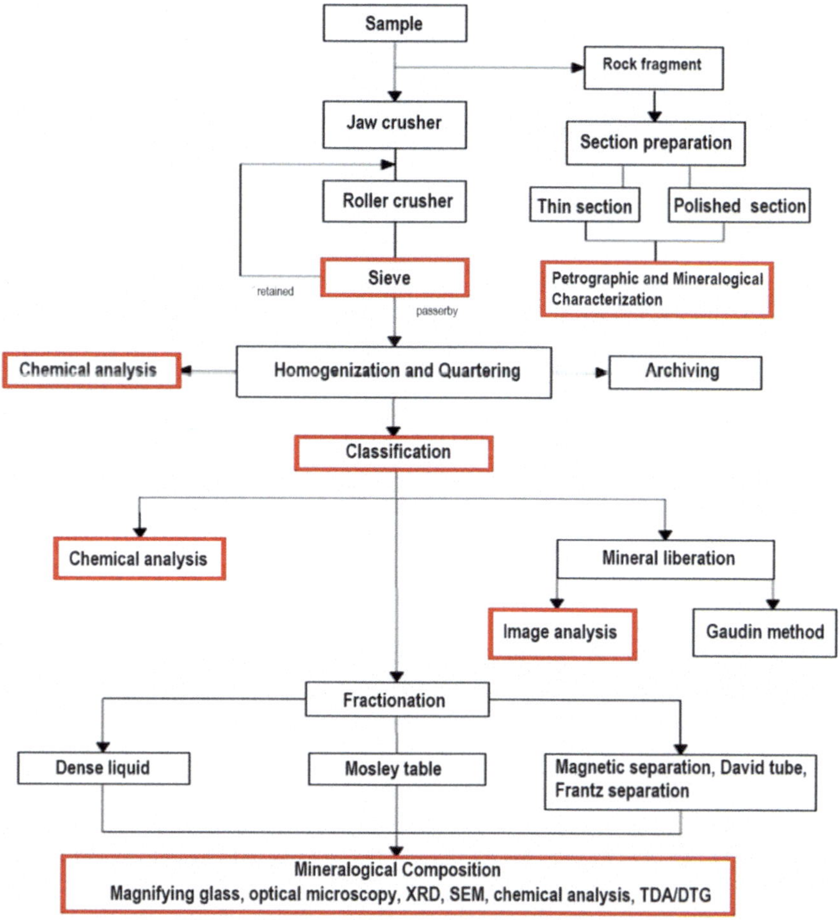

Fig. 1.1 Flowchart of the characterization of mineral raw materials for industrial use (adapted from R. Neumann, C. L. Schneider and A. Alcover Neto CT2004-181-00 Technical Communication prepared for the 4th Edition of the Ore Treatment Book, CETEM, MCT, 2004)

Chapter 2
Chemical and Mineralogical Analysis

2.1 Introduction

The chemical and mineralogical analysis of mineral raw materials is a necessity not always easy to execute, mainly due to the heterogeneous physical nature of the samples. The chemical and mineralogical analysis of ores represents one of the most important pieces of information for the definition of the quality of a raw material. Therefore, chemical and mineralogical analyses are a means and not an end within the industrial process, and the demand and requirements in terms of quality/productivity/technological competitiveness have caused this area to experience exponential growth in recent years.

Chemical and mineralogical analyses are the responsibility of chemists and not engineers, who are only users of chemical laboratories, but some knowledge in this area the engineer must possess to have discernment in the request for analysis and in the interpretation of results. A consistent and accurate exchange of information will facilitate the obtaining of the desired analyses. The engineer, in turn, knowing the analysis technique, will have a clearer notion of the errors and inaccuracies inherent to the analysis, and where the consistency of the results lies in order to use the information obtained appropriately.

The choice of an analysis method should consider the ***principle*** of the method (detection of atoms, ions, compounds), ***what*** is desired to analyze, ***what precision*** (reproducibility) is necessary, ***detection limit*** (smaller elements), ***availability of time and resources*** (**standards!**).

Any analysis technique should consider not only the small aliquot that is submitted to the analysis, but its representativeness of the raw material of interest. The sample sent for analysis must comply with very well defined and described sampling criteria in the literature. When it comes to evaluating raw materials in the mineral industry, a few grams represent tons of material.

Analyses can be **qualitative** and/or **quantitative**. They can be done by traditional methods called **wet route** or classical or by modern **instrumental techniques**, which

C. P. Bergmann, *Techniques for Analysis of Mineral Raw Materials*, Topics in Mining, Metallurgy and Materials Engineering, https://doi.org/10.1007/978-3-031-58985-0_2

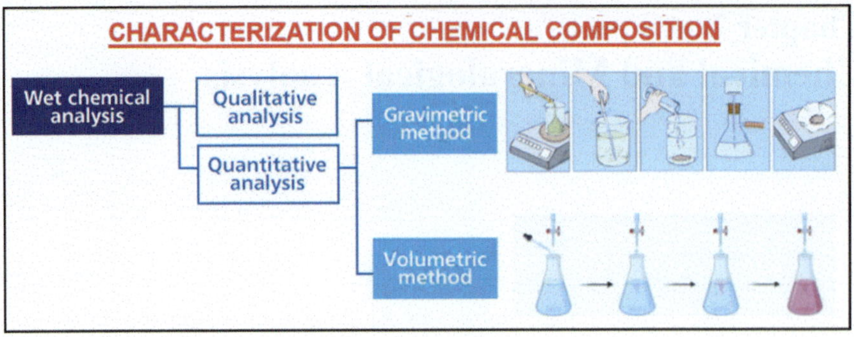

Fig. 2.1 Methods of characterizing the chemical composition of raw materials by wet route

provide quick responses, enabling useful information for research and industrial production.

Until recently, wet route chemical analysis methods were the most common and accurate for determining the chemical composition of a raw material. However, they are generally difficult, time-consuming, expensive, requiring suitable facilities and lengthy training of laboratory technicians. Despite all these difficulties, they are still used in many companies, either because there is no equipment that can replace them, to provide secondary standards for instrumental calibration, as well as to replace equipment during maintenance periods.

Wet route techniques include **titration** (titration of compounds in front of a specific color indicator) **gravimetry** (precipitation of a compound and dosage by the obtained weight) (Fig. 2.1). Trace elements are dosed with great difficulty, with very specific methods. For less precise routine analyses, the duration can be 4 days, while precision ones may require more than a week.

2.1.1 Instrumental Analyses

The development of instrumental analysis equipment intensified from the 1980s, based on the speed and accuracy in obtaining and processing information mainly by the phenomenon of **spectroscopy**. This development led to the emergence of analytical methods **instrumental**, also referred to as methods **spectrometric** or **spectrometric**.

Spectroscopy is the part of Analytical Chemistry for the identification of chemical elements and substances, through the transmission, absorption or reflection of radiant energy incident on a sample. Therefore, when the energy of an electromagnetic radiation is transmitted, absorbed or reflected upon contact with the atoms (or molecules) of a sample, the spectroscopic response of its constituents is a unique characteristic.

Spectroscopic analytical techniques can be classified by wavelengths they work with, the detection principle they use, the way they acquire a spectrum, and intensity of the energy sources they employ. Other important characteristics of spectroscopic techniques include spectral bandwidth and linear range. The simplest measure in spectroscopy is the measurement of light absorption, but it can involve the measurement of diffuse or specular reflectance. The sequence of events in spectroscopy is, as a rule, the following:

(i) the light source shines on the sample;
(ii) the sample absorbs part of the radiation;
(iii) the detector detects the amount of light that the sample absorbed;
(iv) the detector converts the amount of light absorbed by the sample into a number.
(v) the numbers are immediately processed, for which, often some treatment is necessary (for example, baseline correction).

This sequence can be preceded by calibration when some standard substance is defined as a base value, by which the measurements of the samples under analysis are referenced.

2.1.1.1 The Electromagnetic Spectrum and Analytical Techniques

Figure 2.2 presents the types of spectra obtained from the decomposition of white light and sunlight. It is possible to identify two types of spectra: the **continuous spectrum** (as in the case of decomposition by a prism of white light); and the **discontinuous spectrum** (as in the case of decomposition by a prism of sunlight), where the spectral lines characteristic of this type of spectrum appear.

The emission of energy by atoms occurs when electrons jump from one energy layer to another. In the case of molecules, the energy emission is based on rotation and vibration movements and on the movement of the electrons of their constituent atoms. The interaction between the **radiation** or **electromagnetic wave** and the matter (atom, molecule or ion) of a sample is analyzed and measured through a device called a **spectroscope**.

There are two major classes of spectroscopic measurements: those obtained by a **single beam or monobeam** (measurement of the intensity of the absolute light) or by **double beam** (relates the intensity of light through two distinct paths, one passing through a standard sample). The result is expressed in the form of a graph of intensity (of the absorbed or emitted energy) as a function of the wavelength—or more commonly the frequency—called spectrum, spectrogram or spectral line width. What will vary is the nature of the measured excitation:

(i) Absorption spectroscopy: analyzes the amount of energy absorbed as a function of the wavelength of the incident radiation (Fig. 2.3);

(ii) Emission spectroscopy: analyzes the amount of energy emitted by a sample as a function of the wavelength of the absorbed radiation. It fundamentally consists of the re-emission of energy previously absorbed by the sample (Fig. 2.3);

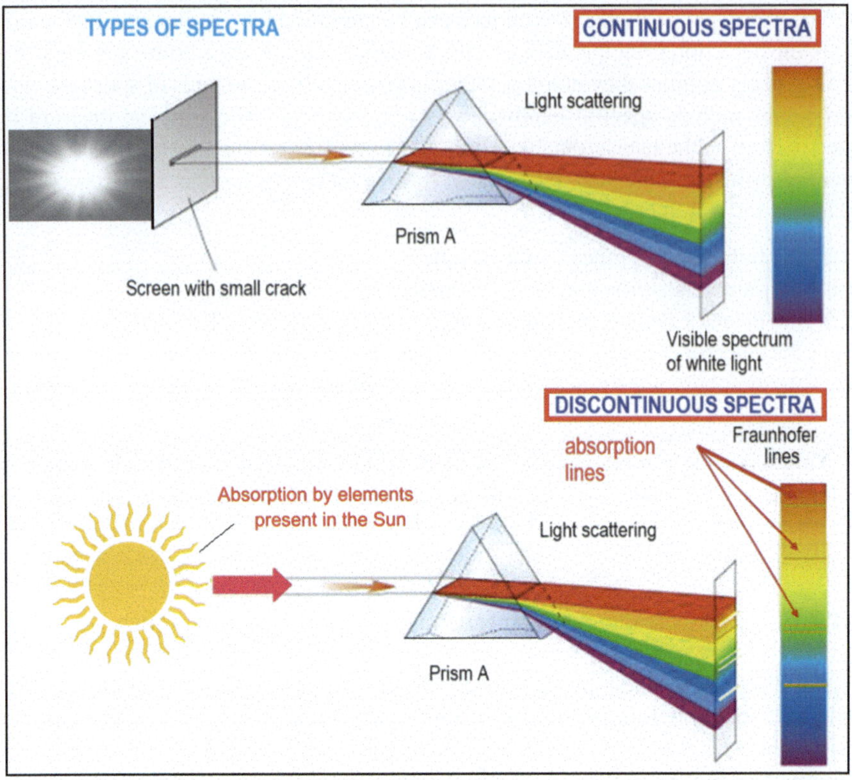

Fig. 2.2 Types of spectra obtained from the decomposition of white light and sunlight. It is possible to identify two types of spectra: the **continuous spectrum** (as in the case of decomposition by a prism of white light); and the **discontinuous spectrum**

(iii) Scattering spectroscopy: analyzes the amount of energy scattered (dispersed) as a function of parameters such as the wavelength, incidence angle and the polarization angle of the incident radiation.

Thus, according to the interaction of the electromagnetic wave with matter, the electromagnetic spectrum can be divided into regions, for better understanding of the fundamentals of analytical instrumental methods, since physical interactions follow distinct mechanisms and provide different types of information, depending on the transitions that can occur (Fig. 2.4).

Depending on the incident energy and the absorbed energy, atoms emit spectra characteristic of their electron distribution in the electrosphere. The distribution of energy levels in the electrosphere is the "identity" of the chemical element interacting with the incident electromagnetic radiation.

Understanding the interactions between matter and energy, and their applications, becomes easier with knowledge of the **Rutherford-Bohr atomic structure model**: in a neutral atom (not excited), electrons occupy as many levels as necessary, starting

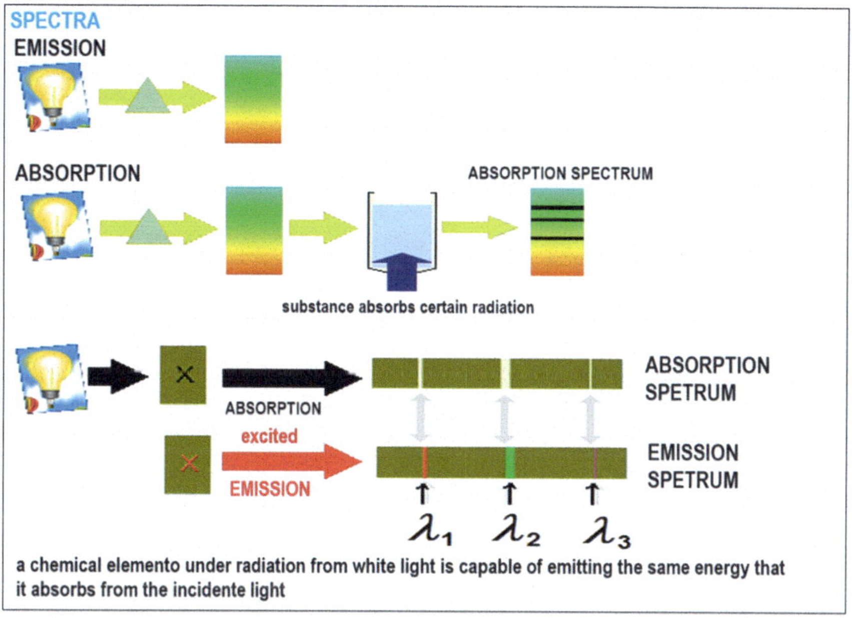

Fig. 2.3 Absorption and emission spectra that occur in the interaction of the electromagnetic wave with matter, which serve as the basis for spectroscopic analysis

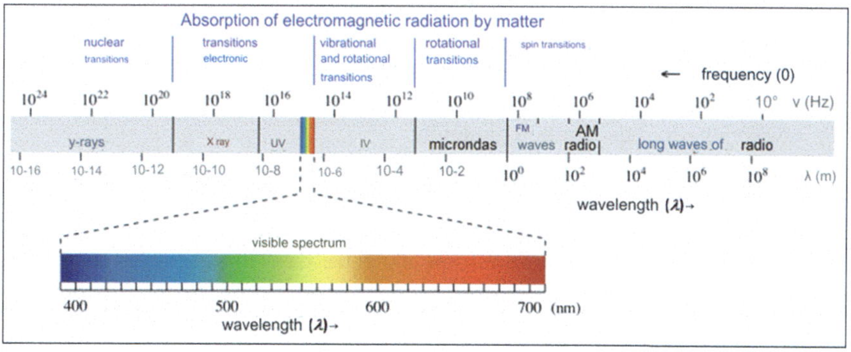

Fig. 2.4 Division of the regions of the electromagnetic spectrum, by wavelength and frequency. In detail, the visible spectrum and forms of absorption of electromagnetic radiation by matter depending on the type of radiation

with the lowest, *1s*, and continuing upwards, according to well-known quantum rules. Sodium, for example, has eleven electrons, distributed as shown in Fig. 2.5.

The electron *3s* is the weakest bound and thus can be easily taken from level *3s* to *3p*, which is an example of electronic excitation. The excited electron has a strong tendency to return to its normal state, the *3s*, and in doing so, emits a *quantum* of radiation (a photon). The emitted photon has a well-defined amount of energy that

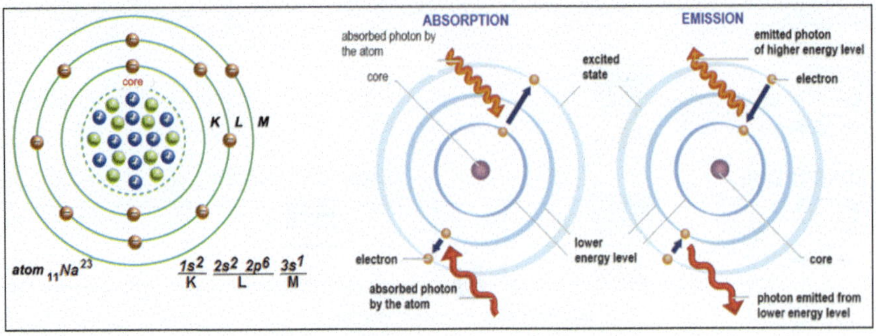

Fig. 2.5 Distribution of electrons in the electron shell of the sodium atom. Absorption and emission of energy by an electron from the electron shell

reflects the energy difference between the energy levels: it is the electromagnetic radiation, characteristic individual of each chemical element.

In the case of sodium, a flame is enough to excite the electron *3s* which, upon returning to its ground state, will release (light) energy that will turn the flame yellow. The quantification of this process is given by Planck's relation (Eq. 2.1):

$$E = \frac{hc}{\lambda} \tag{2.1}$$

where:

E = energy variation (quantum).
h = Planck's constant (6.626×10^{-34} J.s or 4.135×10^{-15} eV.s).
c = speed of light.
λ = wavelength of the emitted radiation.

In the case of sodium, λ corresponds to 5890 Å, a wavelength located in the visible region of the electromagnetic spectrum—and visible to human eyes as yellow color. This simple case, where an outer electron is taken to a higher energy level and then returns, is known as **resonance radiation** and is the basis of the atomic absorption technique. With detectors better than human eyes, it is possible to measure both the energy absorbed by sodium when its electron *3s* moves to *3p* (**Atomic Absorption Spectroscopy**), as well as the energy released when the electron returns to its *3s* position.

Figure 2.6 presents the characteristic spectrum of different chemical elements, which allows their identification in chemical analysis by spectroscopy.

Atomic absorption and emission depend on the source of energization (incident energy). The combustion of air and LPG (Bunsen burner) can reach 1700 °C, air and acetylene generate 2100 to 2400 °C, acetylene and nitrous oxide reach 2600–2800 °C, while the spark of an electric arc can reach 5000 K. The flame of a plasma, with temperatures even above 10,000 K, produces a large number of emission lines, with this energy being sufficient to excite not only atoms, but also ions that reach the plasma.

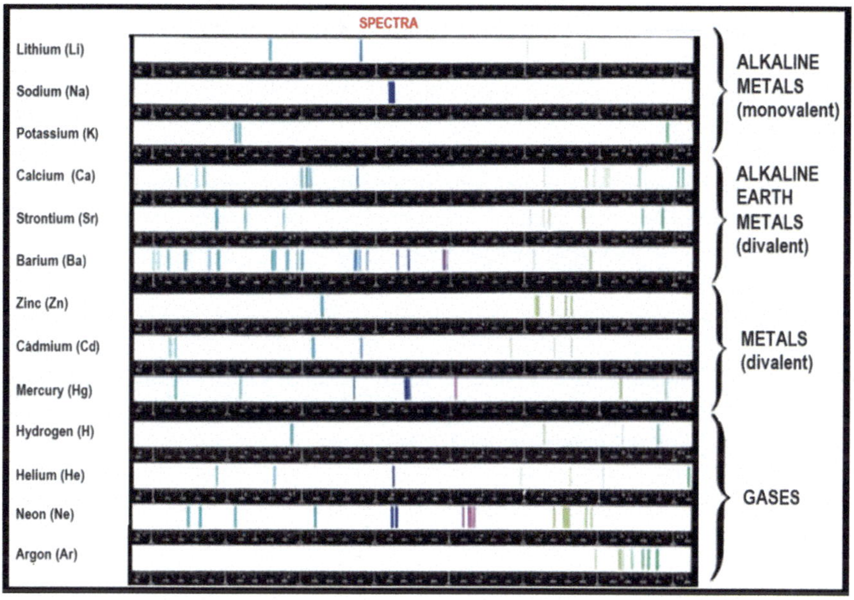

Fig. 2.6 Characteristic spectrum of different chemical elements, which allow them to be identified by spectroscopy

If more energy is provided than necessary to produce resonance, the electron becomes more excited and can be taken to the level *4p*, having more than one path to return to the normal (non-excited) level. With a source of energy even more intense, many electrons can be excited to **various degrees** and the resulting radiation may contain many discrete and reproducible wavelengths, especially in the regions of ultraviolet and visible: this is the basis of **ICP Spectroscopy (Plasma)**.

If the source of excitation is extremely energetic (based on the Planck equation), an inner electron may be completely removed from the atom, and an electron from a higher level will come to fill the gap, releasing the difference in energy between one position and another as a photon. As the energy exchange corresponding to this transition is much greater than in the case of excited electrons, the emitted photons will be of much higher frequency ($E = h\nu$) and, correspondingly, of shorter wavelength ($\nu = c/\lambda$), no longer in the visible region but in the X-ray region. Observing Planck's relation, it can be understood that each transition can be identified by the change in the electron's energy level, as the same transition (for example, *3p* to *2s*) in different atoms will have different distances due to the atomic radii being different. This is the basis of the technique of **X-Ray Fluorescence Spectroscopy**.

2.1.1.2 Interactions with Molecules and Analytical Techniques

As shown in Fig. 2.7, in addition to transitions between electronic levels, transitions due to rotations and vibrations of molecules should also be taken into account. The movement of the atoms that make up the molecules results in molecular rotations and vibrations. As the energies involved in the different forms of rotation are very similar (the difference between two rotational levels is typically around 5×10^{-5} eV), only vibrations are generally considered.

Basically, molecular vibrations can be classified into two types: axial deformation vibrations (*stretching*) and angular deformation (*bending*), as shown in Fig. 2.7. Axial deformations, or stretching, are radial oscillations of the distances between the nuclei while angular deformations involve changes in the angles between the bonds or, as in the mode of out-of-plane asymmetric deformation, changes in the angle between the plane containing the bonds and a reference plane.

When molecular vibrations result in changes in dipole moment, as a result of the variation in the electronic distribution around the bonds, transitions between the vibrational levels can be induced. This is done with the application of electric fields with suitable frequencies (energies). When the oscillating dipole is in phase with the incident electric field, energy transfer from the radiation to the molecule occurs, resulting in a transition.

Two requirements must be observed for a certain radiation to be absorbed by a molecule:

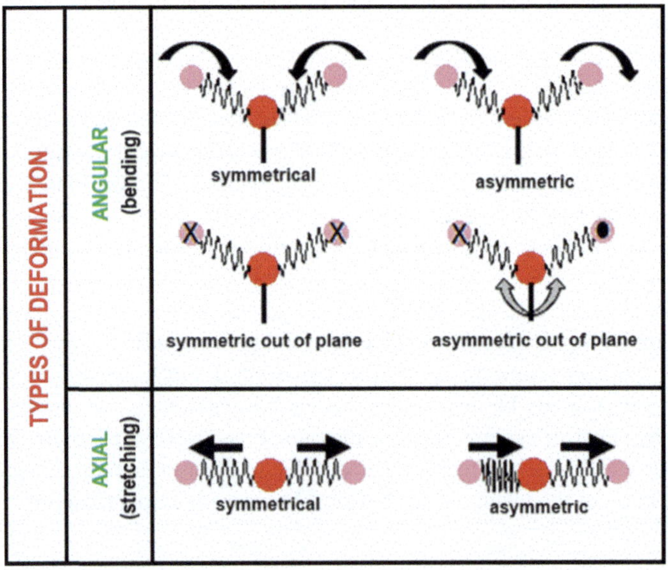

Fig. 2.7 Modes of molecular vibration. The X and ˙signs indicate movements into and out of the plane of the drawing, respectively

i. The incident radiation must be of a frequency equivalent to that rotational or vibrational, electronic or nuclear of the molecule;
ii. The molecule must have a permanent dipole or an induced dipole, that is, there must be some work that the absorbed energy can do.

Analytical techniques such as **Infrared Spectroscopy** and **Raman, Mössbauer and Nuclear Magnetic Resonance Spectroscopy** use this phenomenon to identify chemical bonds present in raw materials under analysis.

2.1.1.3 Application of Instrumental Techniques

The application of one or two instrumental techniques is common in industrial practice. The greatest difficulty in performing analyses for mineral raw materials is the wide variety of contents and materials. Both mineral goods and gangue minerals must be analyzed, in fractions where one predominates over the other, and in fractions where one is concentrated and the other is at trace level. Thus, for each element to be assayed, the contents can vary from ppm to many %, making a single analytical technique insufficient.

In the mining area, AA and XRF have complemented each other, with XRF performing more than 90% of the analyses, when no further information about trace elements is needed. ICP could replace XRF + AA, but it would be too laborious.

The element to be assayed can be a limiting factor for the choice of technique: low Au contents will always be assayed by AA, refractory elements such as Nb, Ta, Zr and Hf will go to XRF or ICP, if they have been properly solubilized, alkalis in AA, anions such as F and Cl in specific electrodes, low contents of Cu, Fe, Zn and Ni in AA, high contents in XRF, and so on.

The **spectrometric** methods used for the chemical and mineralogical analysis of raw materials that will be addressed in this text are: X-ray Fluorescence Spectrometry (XRF), Atomic Absorption (AA), Infrared (IR or FTIR,), Inductively Coupled Plasma Spectrometry (ICP, abbreviation of the English acronym), Raman, Mössbauer, Nuclear Magnetic Resonance (NMR) and X-ray Diffraction.

2.1.1.4 Sample Preparation

The preparation of the raw material is essential in many spectrometric instrumental techniques, such as atomic absorption and ICP, in the dosing of elements in solution. The solubilization of materials can be **total** or **partial**, depending on the purpose of the analysis.

The **total solubilization** is based on two conventional techniques:

(i) **fusion**: it can be alkaline, acidic or oxidizing, with subsequent recovery in aqueous or acidic medium. Classic fusions with sodium and/or potassium carbonate have been replaced by lithium metaborate, lithium fluoride + boric acid, lithium carbonate + boric acid, ammonium fluoride, potassium acid fluoride, boron trioxide, among others. Fusions are usually made in platinum crucibles, to avoid contaminating the samples, but alkaline fusions, such as those of sodium hydroxide or peroxide, should be made in nickel or iron crucibles, as the platinum crucible is seriously attacked. Fusions are normally made on 0.1–1 g of powdered sample and the amount of flux is 4 to 10 times greater than that of the material. The recovery of the fusion can be made with water or diluted acids, depending on the flux and the sample, but the amount of solution ends up being 50–500 ml, which dilutes the sample from 50 to 5000 times.

(ii) **acid attack**: uses hydrofluoric acid, for example, to release constituents of interest from silicate structures. Fluorination can be done in the open, with elimination of the extremely volatile SiF_4, and presence of other acids for solubilization of the other metals, or in small autoclaves that retain including Si. Acid attack is preferable to fusions because they do not load the solutions with viscosity from the reagents, as well as generate smaller dilutions, in addition to being cheaper (they do not require platinum crucibles or expensive reagents, such as sodium peroxide or lithium tetraborate).

The **partial solubilization** aims to selectively solubilize, through acids, only the elements to be dosed or the part of the sample that contains the elements of interest. As an advantage, the solution obtained can have a small volume and better detection limits, does not contain many elements and presents fewer possibilities of contamination. In addition, the methods of partial solubilization are faster and more economical than the total ones. Table 2.1 presents a summary of the acids used in selective partial solubilization.

2.2 X-Ray Fluorescence (XRF)

It is the usual method for determining the chemical composition of raw materials. In this method, a suitably prepared sample is irradiated by an X-ray beam. This beam can displace an electron from the innermost layers of the atom; to replace the missing electron, another electron can jump from one of the outer layers and, in doing so, release energy in the form of fluorescent X-rays, emitted at wavelengths that are characteristic of each element. Figure 2.8 highlights the wavelength range of X-rays, in comparison with that of other electromagnetic radiations. Figure 2.9 summarizes the XRF technique in determining the chemical analysis of raw materials.

The emitted rays have a characteristic and well-defined wavelength for each chemical element that constitutes the irradiated material (qualitative chemical analysis), and it is possible through comparison with standard samples, to establish the proportion of chemical elements present (qualitative and quantitative chemical analysis).

Table 2.1 Summary of the acids used in the selective partial solubilization of mineral raw materials

Acid	Application	Examples	Observations
HCl	Ores and minerals, metal alloys Carbonate rocks easily solubilized even when cold	Iron and manganese oxides and hydroxides dissolve with greater or less ease. Hematite and magnetite dissolve in 15 min	Use of simple containers (glass); acid volatilization is easy; the salts obtained are generally soluble in water. However, silicates, for the most part, insoluble
HNO_3	Minerals such as sulfides and phosphates, metal alloys	Excellent solvent for numerous metals	Due to its oxidizing power, nitric acid dissolves most of the elements in their metallic form
Aqua Regia 3:1 v/v (HCl/HNO_3) $3HCl + HNO_3 = NOCl + 2H_2O + Cl_2 \nearrow$	Aggressive action of nitrosyl chloride and nascent chlorine, to oxides, sulfides and sulfosalts, polymetallic ores and metal alloys	Used in the attack on gold and platinoids	Minerals such as cassiterite, columbite-tantalite, zirconite, barite, chromite, rutile, spinel, and some silicates, among others, are insoluble or difficult to solubilize
HF	High affinity for Be, B, Nb, Ta, Ti, and Zr ions, forming important complexes	Action on silica and silicates	SiF_4, easily removed by heating. Fluorides cause disturbances in analyses, in addition to corroding glass materials
H_2SO_4	Rarely used in raw materials when the final quantification is done by Atomic Absorption	Used in the attack on ores such as sulfides, rare earths, fluorsilicates, and Nb, Ta, Ti, and Zr compounds	Sulfuric acid is the densest of the previous acids, causing a strong signal drop
$HClO_4$	Powerful oxidizer, dissolves refractory materials, but poses a risk of explosion in contact with organic matter	It is most often used as an oxidizer, in small quantities, along with other acids, such as hydrofluoric	Perchloric acid is preferred over sulfuric for two reasons: the resulting salts (chlorides) are more soluble than sulfates and the oxidizing power is greater, aiding in general solubilization

(continued)

Table 2.1 (continued)

Acid	Application	Examples	Observations
H_3PO_4	The action on mineral raw materials is similar to that of sulfuric	Used in the solubilization of rare earths, with the advantage of being a slightly faster solubilization	Phosphates, sulfides, chromite, manganese and iron ores are easily solubilized. Restricted use, as this acid has similar action to other more common acids
Organic acids	Very selective action on certain mineral phases	10%v **acetic acid** is used to dissolve carbonates in ores where fluorite and calcite coexist	Cement factories use this acid in the evaluation of raw material

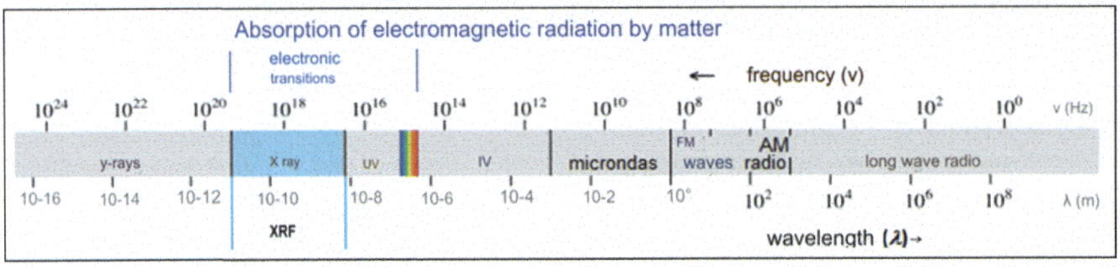

Fig. 2.8 Division of the regions of the electromagnetic spectrum, by wavelength and frequency. In detail, the range of wavelengths of X-rays for XRF

2.2.1 *Principle of the X-Ray Fluorescence Technique*

X-ray fluorescence is based on the principle that the absorption of X-rays by the material causes the excitation of atoms that emit secondary radiation which is called X-ray fluorescence.

X-rays are electromagnetic radiation with a wavelength (λ) between 0.1 and 100 Å. The range of λ useful for X-ray spectroscopy is 0.1 Å (UKα) to 113 Å (BeKα). X-rays propagate at the speed of light *c*, without mass transfer and in a straight line. They are not affected by electric and magnetic fields. They undergo differential scattering and absorption by matter (used in X-ray absorption techniques). They undergo diffraction by crystals (used in X-ray diffraction techniques) and ionize gases (used for detection). They can be produced, through bombardment of materials with electrons, gamma rays, ionized particles and by X-rays.

Figure 2.10 schematically presents the cross-section of a tube used in the generation of X-ray, for materials analysis. The X-rays result from the bombardment of a sealed target (Rh, Cr, W) under high vacuum, by a beam of electrons coming from a heated filament (cathode). The application of a potential difference between cathode and anode causes the emitted electrons to be accelerated, impacting against the target.

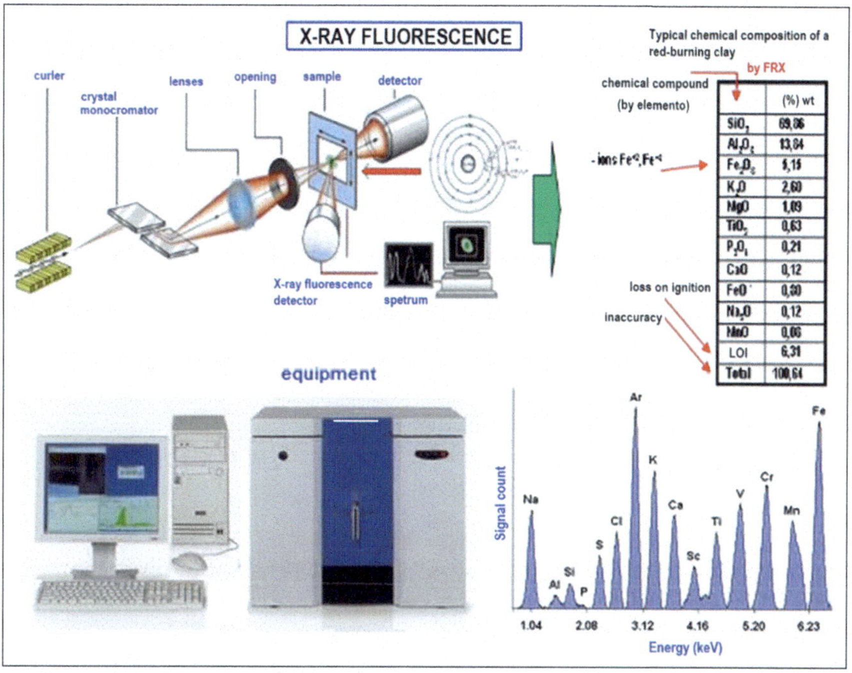

Fig. 2.9 The technique of X-ray fluorescence in the determination of the chemical analysis of raw materials

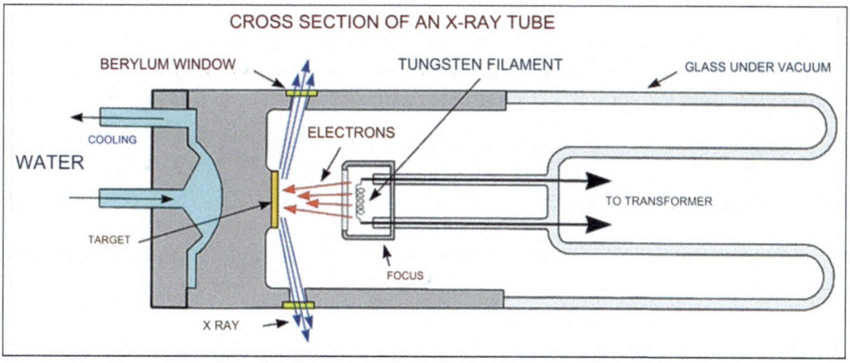

Fig. 2.10 Cross-section of a tube used in the generation of X-ray for materials analysis

When interacting with a material, X-rays typically produce two different phenomena (Fig. 2.11). The first is called **continuous spectrum**, or white radiation or, still, braking radiation (*Bremsstralung*, in German). This radiation is produced by the deceleration of high-energy electrons through inelastic collisions with the electrons of the target atoms. An electron can undergo a series of collisions until it loses

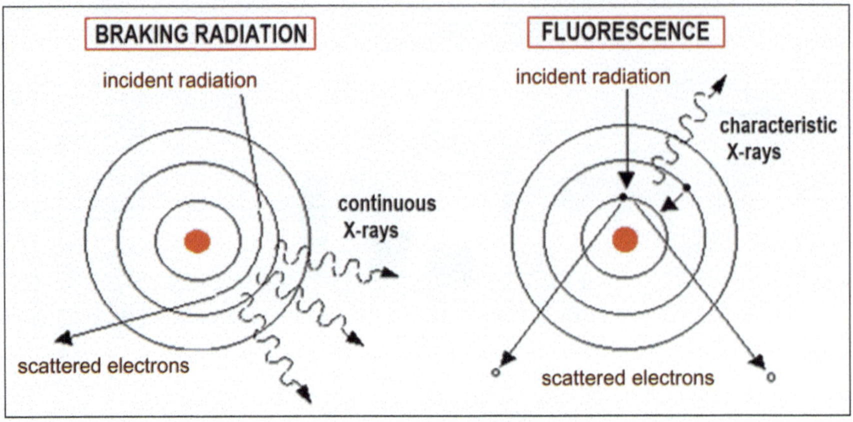

Fig. 2.11 X-ray emission spectrum: continuous spectrum and characteristic spectrum

all its energy. In each collision, it releases the lost energy in the form of photons of electromagnetic radiation. The continuous spectrum depends on the operating conditions of the tube (voltage and current), as well as the target material, according to Eq. 2.2.

$$I \alpha i \, .Z \, . \, V^2 \tag{2.2}$$

where

i = current intensity;

Z = atomic number of the target chemical element

V = voltage of the electric field.

The other phenomenon is the **characteristic spectrum** which is like the "fingerprint of the atom" of the chemical element under X-ray radiation. This phenomenon is therefore the basis of the method of chemical analysis by X-ray fluorescence. It consists of the emission of photons with wavelengths characteristic of each atom, since it depends on the binding energy of the electrons to the nucleus, which is defined by the atomic weight of the chemical element. The emergence of the characteristic spectrum occurs only when the incident photons have a minimum energy, above the absorption band of the material in question.

The characteristic spectrum shows the electronic rearrangement resulting from the fall of the excited target electrons from the fundamental level. The emission lines, discrete and superimposed on each other, translate the point at which the energy was sufficient to completely remove one of the target electrons and show that another electron took its place, emitting a **photon of X-ray radiation, with a wavelength corresponding to the energy levels involved**, and therefore, characteristic of the target element. Therefore, when an X-ray photon with sufficiently high energy hits an atom, it can cause the ejection of electrons (photoejection) from the innermost

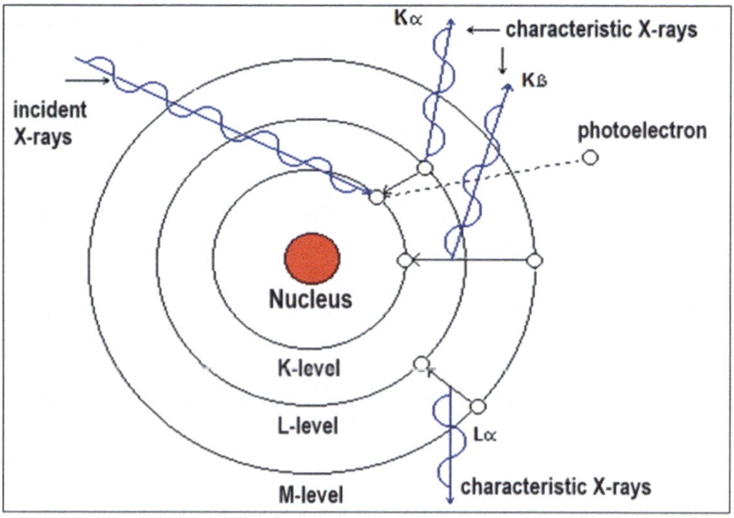

Fig. 2.12 The characteristic spectrum of an atom under X-ray radiation. Electronic rearrangement resulting from the fall of the excited target electrons from the fundamental level

layers (K, L and M). Electrons from higher layers (higher energy) decay to fill the gap formed, releasing the energy difference in the form of an X-ray photon (Fig. 2.12).

Figure 2.13 presents the characteristic spectrum and continuous spectrum, with the wavelengths corresponding to the energy levels involved. For the characteristic spectrum, the Greek letter indicates the layers involved, while the numbers refer to the sub-layers. The emission of the characteristic spectrum (e.g. Kα) always occurs in pairs ($K\alpha_1$ and $K\alpha_2$), as indicated in Fig. 2.14.

The relationship between the frequencies of the α spectra of the K and L series of characteristic X-rays of various elements, and the atomic number, Z, of these elements was defined by the English scientist Henry Moseley, in 1913. To do so, Moseley measured the frequencies of the spectral lines of the characteristic X-rays of about 40 elements. From the graph of the square root of the frequency (f) *versus* the atomic number Z of the element, Moseley obtained the relationship given by Eq. 2.3.

$$f^{1/2} = A_n(Z - b) \tag{2.3}$$

where

Z = atomic number;

A_n and b = constants that depend on the spectral line.

Therefore, the square root of the frequency of the X-ray spectra of any element is a simple linear function of the atomic number of that element. Moseley's law allows in a simple and direct way, to evaluate and identify the chemical element by its atomic number, in the analysis of materials by X-ray fluorescence (Fig. 2.15).

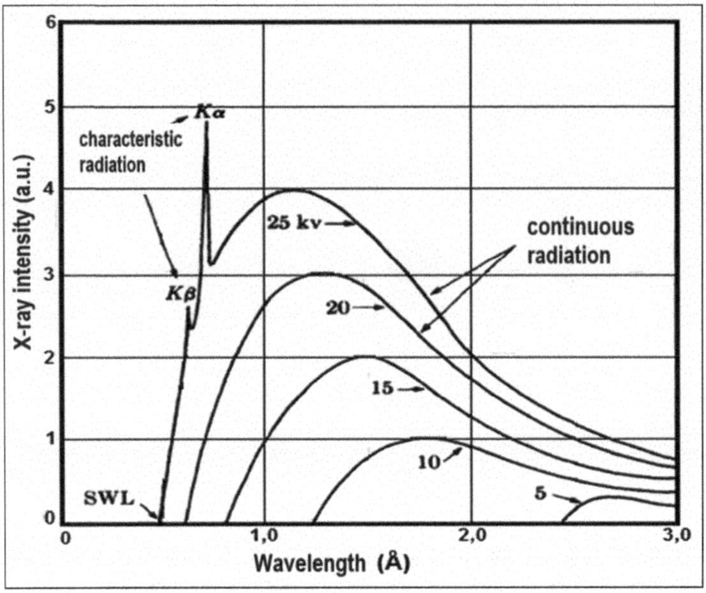

Fig. 2.13 Lines of the characteristic spectrum and the continuous spectrum, with the wavelengths corresponding to the energy levels involved in the interaction of X-rays with a chemical element

2.2.2 *Applications of X-Ray Fluorescence*

X-ray fluorescence is intended to analyze solid, liquid and gaseous samples, detecting elements with an atomic number greater than 11 up to uranium (the last natural element existing in the Periodic Table), in concentration ranging from traces to its maximum concentration in the analyzed sample.

XRF is a widely used and suitable technique in various industries, for many advantages: it is fast, it is non-destructive and does not alter the samples, it can be done on solids (powders, metals, ceramics, plastics) or liquids, it can be a qualitative analysis (scanning), semi-quantitative or quantitative, almost all elements of the Periodic Table can be dosed (from Be to U, therefore, does not detect Li, He and H). The scanning time (from $_5$Be to $_{92}$U) is typically 150 s for qualitative analysis. Anions can be dosed, accuracy and reproducibility are high and concentration ranges are wide (from ppm to close to 100%).

The improvement of instrumental conditions brought small and continuous advances, that is, better resolution and better acuity. Using an old concept (instrumental sensitivity) in a new way (sensitivity as instrumental calibration), it is possible today to analyze 72 elements chemicals, in 15 min, without the aid of standards similar to the samples, without raising calibration curves. Metals, rocks and ores can be analyzed, without standards and in any sample size. They are reliable semi-quantitative analyses, sometimes being compared to quantitative ones. The samples can be in the most different shapes and dimensions, as suggested by Fig. 2.16.

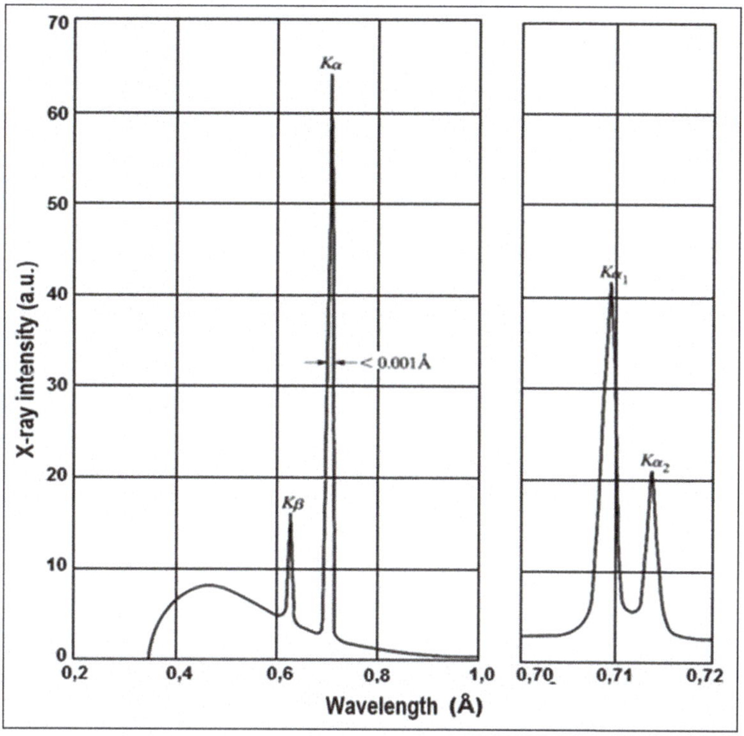

Fig. 2.14 Characteristic spectrum, where the Greek letter indicates the layers involved, while the numbers refer to the sub-layers. The emission of the characteristic spectrum (e.g. Kα) always occurs in pairs ($K\alpha_1$ and $K\alpha_2$)

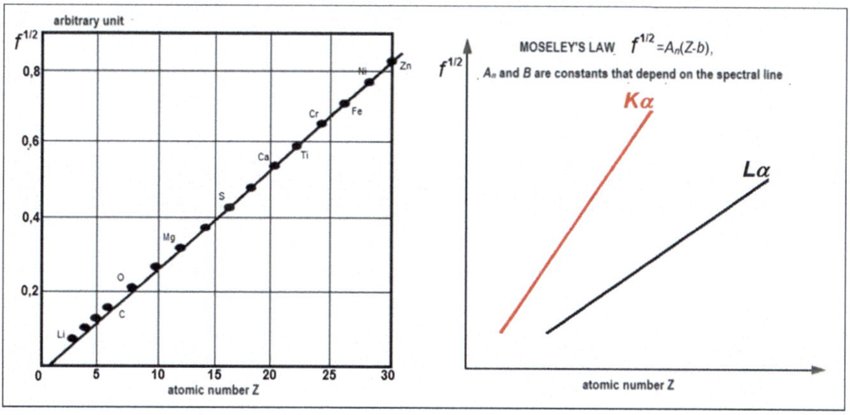

Fig. 2.15 Moseley's Law: the square root of the frequency of the X-ray spectra of any element is a simple linear function of the atomic number of that element

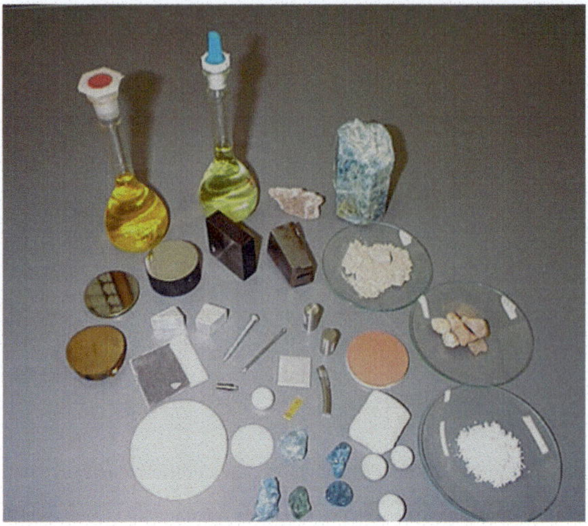

Fig. 2.16 Sample for chemical analysis by X-ray fluorescence

The diameter of the samples can vary from 5 to 30 mm. The detection limits are of the order of ppm's, depending on the sample, the element and the conditions of analysis.

XRF is a non-destructive analysis, with the radiation spectra not being affected by chemical bonds. As it is an elemental chemical analysis, it does not provide any structural information (crystallographic, nature of the compound, molecular) of the sample. Normally, the results of XRF are provided in % by weight of the cation, or % in moles. This way of expressing the results can lead to inconsistencies in interpretation (the standard is to generate reports with results expressed in terms of oxides of the cations found). For example, in a raw material containing various compounds of Ca ($CaCO_3$, $Ca(OH)_3$ and CaO. The result would normally be expressed as being 100%Ca or 100% CaO.

Description: DCP03108

Therefore, as far as it is possible to know in advance, it is necessary to provide data to the software of the equipment, so that the final result takes into account the existence of other compounds (or even organic matter). Thus, as a rule, loss on ignition at 1000 °C is performed, prior to the analysis.

X-ray detectors have the function of converting the energy of the photons emitted by the sample into electrical pulses. The detectors work with a photoionization process where the interaction between the energy of the photons and the detector material produces a certain number of electrons. The current produced is converted into pulses by a capacitor and a resistor, so that a pulse is produced for each photon.

There are two types of analyses: WDS spectrometry (wavelength dispersion) and EDS (energy dispersion). Figure 2.17 presents both XRF-WDS and XRF-EDS techniques.

In WDS spectrometry, the various X-ray lines emitted from the surface of the sample are dispersed through the diffraction of specific crystals for wavelength

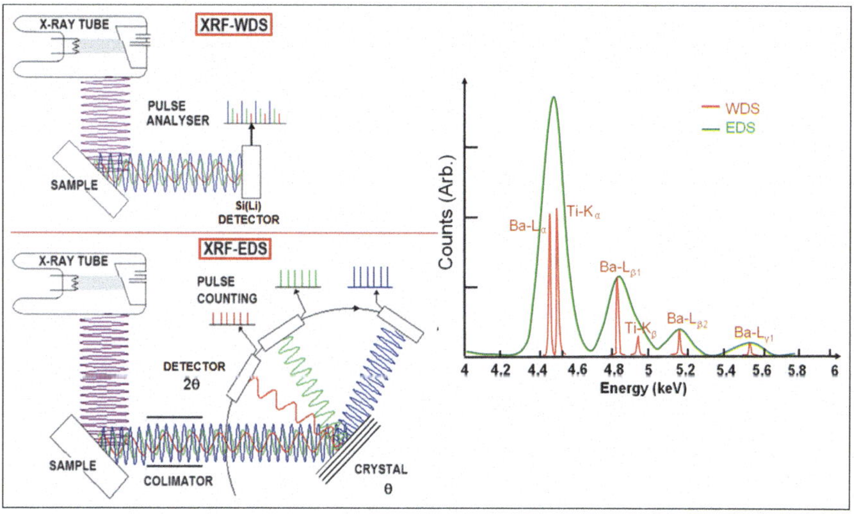

Fig. 2.17 WDS XRF Spectrometry (wavelength dispersion) and EDS XRF (energy dispersion). Adapted from https://www.globalsino.com/EM/page3110.html

ranges. In this way, the detector receives only a certain wavelength at a time. The crystal and the detector are programmed to rotate synchronously at angles of θ (crystal) and 2θ (detector).

In EDS spectrometry, the primary X-ray beam incident on the surface of the sample excites the formation of several spectral lines (wavelengths), which together impinge on the detector. The detector generates current pulses, with the signal height of these pulses, proportional to the energy (concentration). These pulses are then electronically separated through a pulse analyzer.

The main differences between the two X-ray techniques, WDS and EDS, are based on the source of excitation, the number of elements analyzable, the speed of analysis and its price. Both equipment, in principle, can analyze almost all elements between F and U. In XRF-EDS the detector captures the entire spectrum simultaneously, while in WDS, the detector captures the characteristic lines that were separated by the analyzing crystal.

WDS can be sequential (one element analyzed at a time by a single detection system) or simultaneous (several channels present in the system). EDS was and is more used in microprobes coupled to electronic scanning microscopes, where qualitative or semi-quantitative analysis is more important than quantification, but it has been gaining space as an independent analytical instrument. The cost of the two pieces of equipment is quite different, due to the embedded components and potential of the technique.

For quantitative analyses where greater flexibility is desired and where speed is not critical, with a justifiable moderate initial cost, sequential WDS is the ideal equipment. Table 2.2 presents a comparison between the characteristics of both pieces of equipment.

Table 2.2 Comparison between WDS and EDS techniques for material analysis by X-ray fluorescence

WDS Spectrometry	EDS Spectrometry
Better resolution	Simplicity of the instrument—no moving parts
Higher individual intensities can be measured, as only a small part of the spectrum falls on the detector	Compact system
Lower detection limits	Simultaneous recording of the entire X-ray spectrum
Better sensitivities in light elements (below $_{11}$Na!)	Measurements can be performed in short times (±40 s)
	Low X-ray power (±25 W)
	Lower price

2.2.3 *Sample Preparation for XRF*

A general rule says that the ideal sample for XRF analysis is flat, homogeneous, and thick. The simplest sample preparation is pressing, e.g., 10 g of sample mixed with 1.5 g of wax after homogenization. This preparation is used to quantify trace elements in rocks, soils, sediments, for applications in Environmental Geochemistry, exploration, and geochemical mapping. The determination of major and minor elements in tablets produces results of low precision and accuracy, but this high uncertainty is still sufficient for some applications (for example, geochemical mapping).

Homogeneity can also be achieved with the preparation of fused tablets. Solid samples that show segregation (typically powders and particulates) are ground to granulometries smaller than 200# (74 μm). They are then pressed into the form of a tablet, using a binder (boric acid, corn starch) and fused. The fusion is carried out with fluxes such as lithium tetraborate (Li_2 B_4 O_7) pure or mixed with lithium metaborate ($LiBO_2$), in a sample: flux 1:2 to 1:10. The fusion is carried out in special alloy crucibles (Pt with 5% Au) and the melt is poured into molds of the same material. The controlled speed cooling produces circular glasses, in the form of a disc. Fusion is the most recommended method to prepare the specimens to quantify larger and smaller elements in rocks and other mineral raw materials.

Liquid samples can be analyzed by the filter paper method, where the solution to be analyzed is placed on a filter paper and left to dry. The paper containing the precipitated salts is then analyzed, using special sample holders (with Mylar window, as per Fig. 2.18).

For quantitative analysis, due to differential absorption effects of the atoms present in the sample, standards composed of the same elements of the sample should be used. The standards can be constructed with any compound that contains the desired atoms, in known concentrations.

Fig. 2.18 Sample holder for analysis of liquid samples (Mylar window)

2.2.4 *Sources of Error in XRF*

The most important source of deviations in XRF is the variation due to the weight of different matrices: different backgrounds (*bg*) are derived from the absorption level of the set (matrix) of elements present in the samples and can significantly alter the results. For quali-quantitative analysis without standards, empirical methods and algorithms are used for matrix effect correction, through appropriate software (Fig. 2.19). What is common to do in XRF is to establish peak reading and *bg* to compensate for these variations.

In pressed powders, a significant problem is the granulometric effect, which distorts the values obtained by "shadow" effect, that is, coarse grains casting shadow on fine ones. The penetration of radiation is small, and the irregular distribution of grains in different layers can also bring deviations in the analyses.

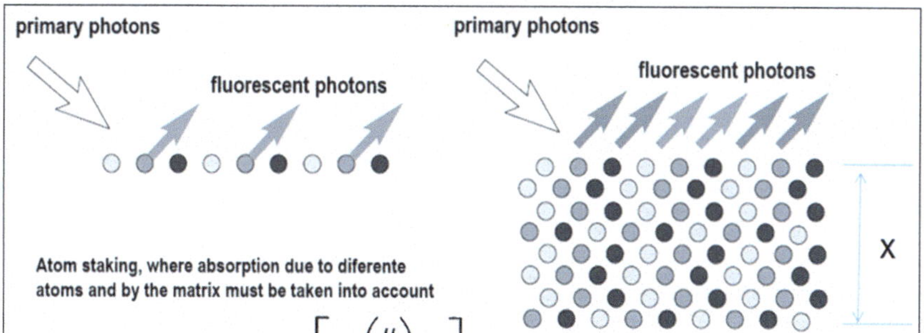

$$I(\lambda) = I_0(\lambda)\exp\left[-\left(\frac{\mu}{\rho}\right)\rho t\right]$$

Fig. 2.19 Matrix effect: absorption of the set of elements presents in the sample. Use of algorithm for the correction of matrix effect

In addition to the mentioned deviations (granulometry and matrix effect), there may occur errors due to spectral interferences. The presence of interfering elements causes signal increase due to total or partial peak overlap (overlap or wings), but as all the peaks of all the elements are tabulated, consulting the table that accompanies the equipment easily reveals the possible interferers. Some examples of spectral interferences are Mn Kβ and Fe Kα, Pb Lα and As Kα, V Kβ and Cr Kα, Zr Kβ and Mo Kα, Y Kβ and Nb Kα, Pb Lβ and Bi Lβ over Th Lα, and Rb Kβ and Y Kα. Some of these interferences can be eliminated by changing the crystal, for example, from LiF 200 to LiF 220 or 420, improving the resolution and separating the two peaks more. More recently, the use of special software can identify and minimize the presence of these interferences in XRF analyses.

2.3 Atomic Absorption Spectrometry

Atomic absorption spectrometry (AAS) is a chemical analysis method for qualitatively/quantitatively determining the presence of metals in samples, based on the phenomenon that establishes that atoms Books in a stable state can absorb light at a certain wavelength. Absorption is specific to each element. It is therefore an elementary analytical technique, with high selectivity. Figure 2.20 presents the division of the regions of the electromagnetic spectrum, by wavelength and frequency. In detail, the range of wavelengths in the ultraviolet and visible regions, where the atomic absorption technique operates.

Atomic absorption spectrometry emerged in the 1950s, replacing titration and gravimetry operations of wet chemical analysis, providing more speed in chemical analyses. Currently, AAS is a well-established and robust technique for laboratories involved with large-scale chemical analyses.

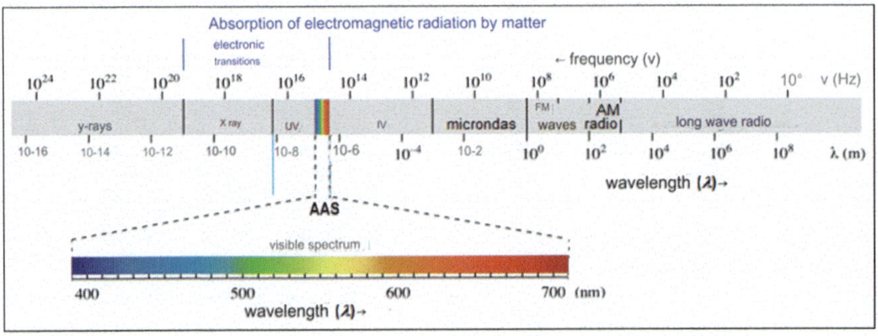

Fig. 2.20 Division of the regions of the electromagnetic spectrum, by wavelength and frequency. In detail, the range of wavelengths in the ultraviolet and visible regions, where the atomic absorption technique operates

The main advantages of this technique are: high specificity; high sensitivity; robustness; low detection limits for various elements in different samples and relatively low cost. AAS is used exclusively for the analysis of metallic elements, preferably in low concentration (<1%). It is widely used in the mining and metallurgical industries, and in environmental analyses, among others.

2.3.1 Principle of the Atomic Absorption Technique

This technique is based on the absorption of characteristic radiant energy, in the ultraviolet and visible regions, by neutral and non-excited atoms in a gaseous state. An AAS device is basically constituted by a system of nebulization and vaporization of solutions, an energy source (flame) for atomization and excitation of the elements, a source of resonance line emission, and a signal detection system. Figure 2.21 schematically presents the atomic absorption spectrometry analysis technique.

The most used source for atomic absorption measurements is a hollow cathode lamp. It consists of a tungsten anode and a cylindrical cathode supported in a glass tube that contains inert gas, such as argon. The cathode is made with the element to be analyzed. The heat to gasify the sample is generated from a flame or graphite furnace. Flame AAS can only analyze solutions, while furnace AAS can analyze solutions and solid samples. A flame atomizer consists of a nebulizer that transforms the sample into an aerosol that feeds the burner.

Normally, the sample is subjected to AAS in solution. The element to be analyzed will be in the form of a cation (anions are not detected in AAS), will be in an aerosol after nebulization and will be dissociated and vaporized due to the temperature of the flame, becoming an atom in the fundamental state. A small portion of the atoms present in the flame will be excited by it and will become cations which, when returning to the fundamental state will produce energy, like Na heated by a flame, emitting a characteristic radiation (in this situation, AAS can be used as a flame photometer, without using the hollow cathode lamp).

The large portion of atoms that did not ionize is susceptible to another source of energy present in the system, which is the light beam emitted by the hollow cathode lamp, whose cathode is of the same metal that one wants to dose and, therefore, of λ identical to those allowed to the element of interest. The atom **absorbs** this energy.

Quantitative analysis is based on the **Beer-Lambert's Law** which quantitatively determines the concentration of substances in solution that absorb radiation (Eq. 2.4).

$$A = -\log_{10}\left(\frac{I_o}{I}\right) = \varepsilon.c.L \qquad (2.4)$$

where

A = *absorbance* measured;

I_0 = intensity of the light incident at a given wavelength;

I = intensity transmitted by the sample;

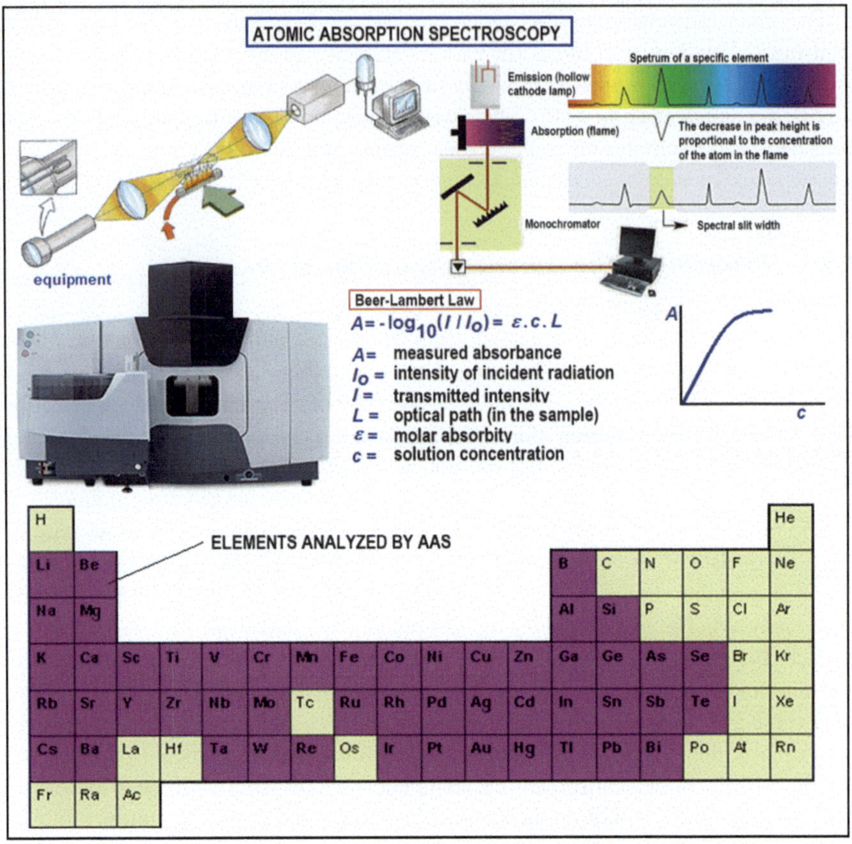

Fig. 2.21 Characteristics of the atomic absorption spectroscopy analysis technique

L = optical path through the sample (distance that the light traveled in the sample);
ε = molar absorptivity (which varies from substance to substance);
c = *concentration* of the substance in the solution (mol/L).

Absorbance and ε are often defined in terms of the natural logarithm instead of the logarithm base 10. The concentration can also be given in (g/L) where instead of ε is used α (absorptivity), where there is a relationship between ε and α given by Eq. 2.5.

$$\varepsilon = \alpha \ . \ MM \tag{2.5}$$

where MM is the molecular mass of the substance analyzed.

Deviations can occur due to the interactions of the absorbing centers and the variation of the refractive index. By the Beer-Lambert Law, the absorbing centers have no interactions with each other or with other species present in the solution, limiting its application to diluted solutions ($<10^{-2}$). This interaction alters the charge distribution in the absorbing species, modifying the energy required for its excitation, therefore the position, shape, and height of the absorption band may undergo changes.

Another real deviation from the Beer-Lambert Law is the possibility of there being a variation of the refractive index n of the solution with the concentration. This arises from the fact that ε depends on the refractive index of the solution. For solutions of low concentrations, n is constant, but it can vary considerably for solutions with higher concentrations. Deviations above 0.01% are only observed for solutions whose concentrations are above 10^{-2}.

The AAS technique uses a narrow radiation source, necessary to reduce interference from other elements and background noise (background). For this, two types of sources are alternatively used: (i) hollow cathode lamp and (ii) discharge lamp. Flameless atomization is only used for elements that emit vapors at room temperature, such as the metal mercury.

Figure 2.22 presents the operating characteristics of these lamps. A **hollow cathode** lamp should produce the emission of lines from the cathode element, which should be a good electrical conductor, relatively volatile and conformable in the form of a filament. The lamp contains an inert gas (like argon or neon) that when heated and excited, its molecules when impacting against the cathode surface knock atoms off it. The repeated bombardment of the metal atoms by the gas causes their excitation. The excited electrons when returning to the normal state produce specific line emissions.

Discharge lamps contain a salt of the metal of interest in a sealed quartz tube full of an inert gas. A radiofrequency (RF) field is used to excite the gas and causes the metal to be ionized. Intense light is emitted at a ratio of 10 to 100 times more than hollow cathode lamps, but in a not so stable way.

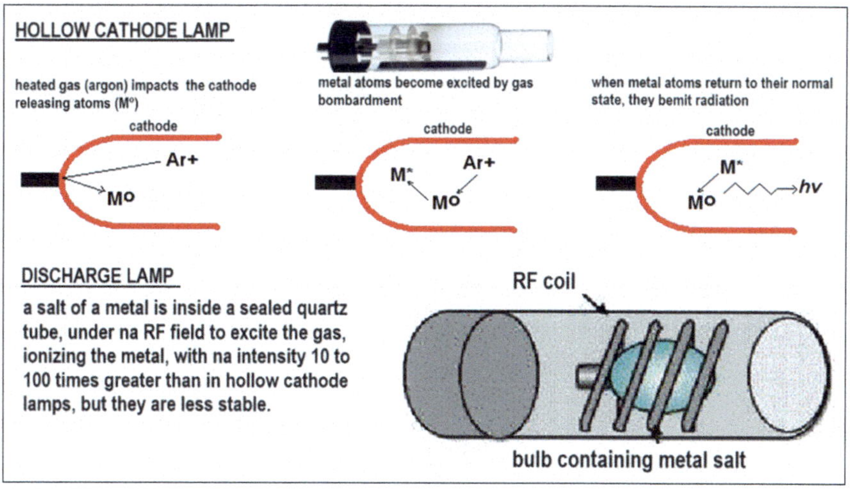

Fig. 2.22 Lamps used as a radiation source in atomic absorption spectrometry

Two **atomization** techniques are used in atomic absorption spectrometry: (i) by flame (used for liquids and gases); (ii) by graphite furnace (used for liquids and solids). Figure 2.23 presents a comparison between the atomization techniques used in AAS.

A **flame** atomizer usually has a long and narrow burner that serves as a path for the sample. The sample is introduced via aspiration. The nebulizer controls the sample flow, producing the mixture between the oxidant (air or N_2O) and the fuel, before

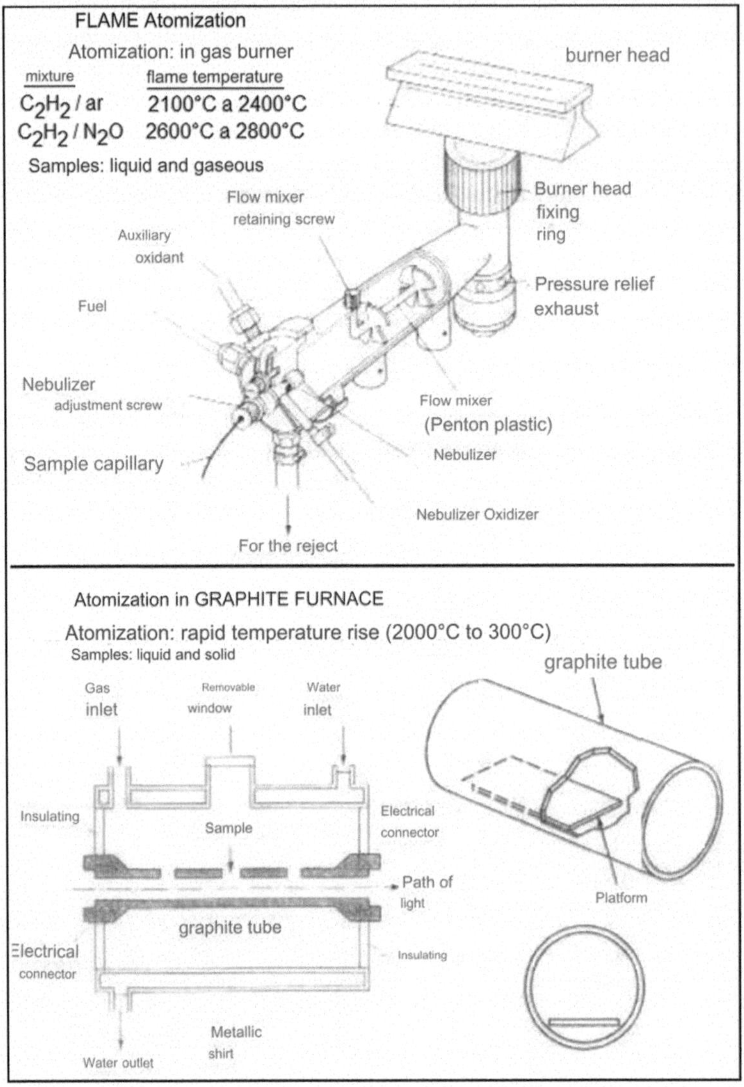

Fig. 2.23 Comparison between the atomization techniques used in AAS

entering the flame. The most common fuel is acetylene, with flame temperature between 2100 and 2400 °C when burning with air and between 2600 and 2800 °C when burning with N_2O (which tends to produce a flame with more noise).

In **graphite furnace** atomization, the sample is placed in a furnace that is electrically heated. Atomization occurs after a rapid temperature increase from 2000 to 3000 °C for a few seconds, when data is collected. Usually, argon is used as purge gas. With this technique, solid samples can also be analyzed.

The absorption of energy can be measured by a detector placed in the path of the beam that reads the amount of energy emitted by the lamp as only water is aspirated and when atoms of the element to be dosed absorb the energy of the light beam. This light absorption is initially measured in units of light absorption, or transmittance ($T - I/I_o$), a unit of measure that is the complement of absorbance ($\%T = 100 - \%A$).

Light absorption is proportional to the number of atoms present, and maintains a practically linear correlation with concentration, for low concentrations. Absorbance is the term used in practice, and the **law of Beer-Lambert** governs this relationship. Thus, the equipment can be calibrated with solutions of known contents, and the interpolation of unknown samples will provide the concentration.

2.3.2 Applications of AAS

Atomic absorption spectrometry is efficient for the determination of the concentration of about 50 metals, in the ppm range, some with a higher degree of difficulty, others with simple and well-known methods. In recent years, it has been superseded by ICP, which is slightly more expensive, but with much more sensitivity. Both for prospecting, characterization, cubing, monitoring of mining, and beneficiation by cyanide leaching, atomic absorption spectroscopy presents sufficient sensitivity.

2.3.3 Sample Preparation for AAS

The samples must be in solution, or at least, the metals to be dosed must have been solubilized (see item 2.1.1.4). Leaching with acids or mixtures of acids are sufficient to put in solution metallic alloys and easily soluble minerals, such as carbonates, sulfides, iron and manganese oxides.

Fusions also lead to the complete dissolution of minerals, but are generally avoided because they make the solutions loaded with salts, which can cause the nebulizer to clog and cause interferences.

2.3.4 Sources of Error in AAS

The main sources of error (or interferences) in AAS can be divided into (see Fig. 2.20):

(i) **spectral**: rare in AAS, since the range of λ of resonance radiations is narrow. The way to circumvent them is by changing the wavelength.

(ii) **chemical**: they are the most common, and occur either by ionization in the flame above the expected, or formation of refractory oxides, formation of high melting point salts, or absorption by molecules coexisting with the atoms in the flame.

(iii) **physical**: different viscosities of solutions, either by the amount of dissolved salts, or by the amount and type of acids used, causing different flow rates and, consequently, different nebulization and vaporization rates. In the preparation of samples by acid attack, the action of the acid on the signal of the element to be analyzed must be considered. HCl is the most used, as the signal reduction it causes is smaller. The graph in Fig. 2.24 shows a comparison of the reduction in absorbances generated by 10 ppm of Ti in AAS, depending on the acid used in the sample preparation.

One way to visualize and circumvent interferences is the method of additions, when known amounts of the element to be analyzed are added to some aliquots of a sample, and their measurements are absorbances, as shown in Fig. 2.24. Samples without interferences will show a line parallel to the aqueous standards, while samples with interferences (whatever the cause) will show disagreement.

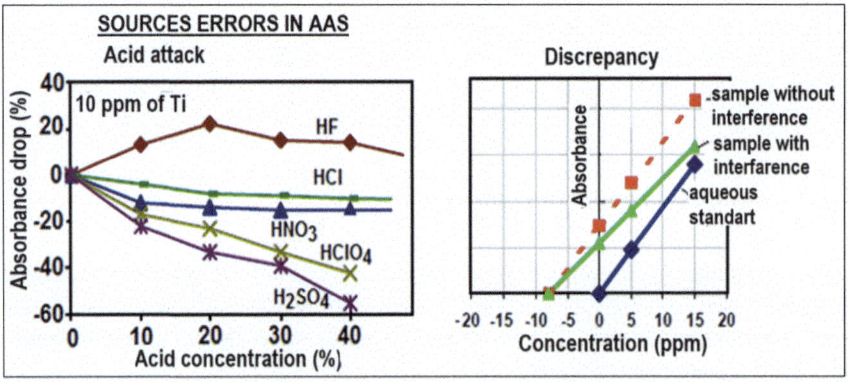

Fig. 2.24 Sources of error in atomic absorption spectroscopy analysis

2.4 UV/Visible Spectroscopy

The **ultraviolet visible spectroscopy** (UV–VIS) involves the spectroscopy of photons (spectrophotometry). This technique uses light in the range of visible, of ultraviolet (UV) near and of near infrared. In these energy ranges, the molecules undergo molecular electronic transitions. Figure 2.25 presents the division of the regions of the electromagnetic spectrum, by wavelength and frequency. In detail, the range of wavelengths in the regions of ultraviolet and visible.

UV–VIS spectra generally only present a few broad absorption bands compared to techniques such as infrared spectroscopy that produces many bands. UV–VIS spectroscopy provides only a few qualitative information for identification and characterization of compounds, but it is an extremely useful technique ın the study of substances that promote color.

To compare the observed color with the obtained spectrum, the correlation with the absorbed color is used. The observed color is the complement of the absorbed, as shown in Fig. 2.26.

2.4.1 Principle of UV–VIS Spectroscopy

In general, when using radiation in the UV–Visible range, the absorption of radiation by the molecules of chemical compounds is measured. The absorption of energy is quantized and causes the passage of electrons from orbitals of the ground state to orbitals of higher energy in a more excited state. When this electron returns to its ground state, it releases exactly the same energy that was absorbed. This energy, when in the visible region of the electromagnetic spectrum, provides the color.

The relationship between the energy absorbed in an electronic transition and the frequency (v), the wavelength (λ) is the same as in Planck's Equation (Eq. 2.1),

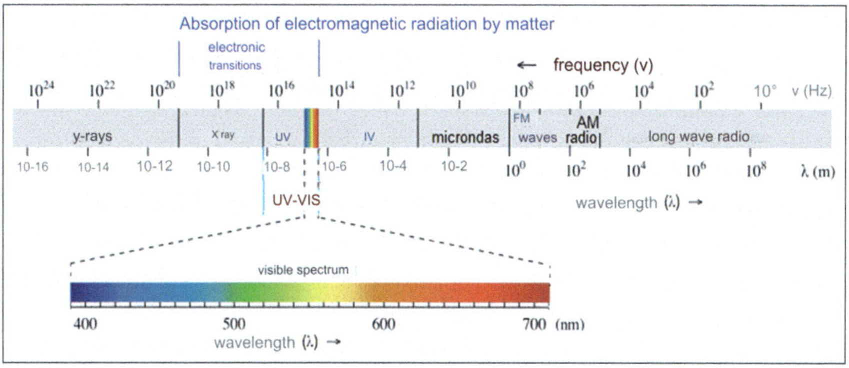

Fig. 2.25 Division of the regions of the electromagnetic spectrum, by wavelength and frequency. In detail, the range of wavelengths in the regions of the ultraviolet and visible

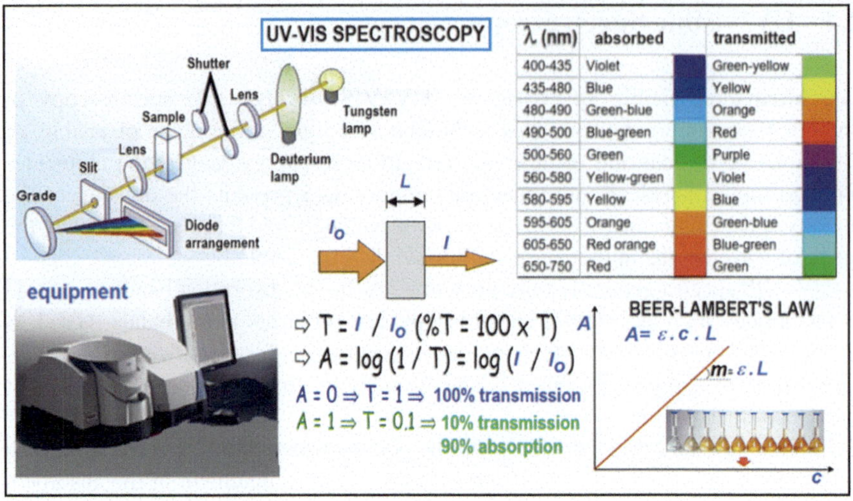

Fig. 2.26 Characteristics of the UV–VIS spectroscopy analysis technique

given by ($E = h.v$ or $E = h.c/\lambda$). E is the energy absorbed by the molecule in the electronic transition between a lower energy state (ground state) and a higher energy state (excited state).

A spectrophotometer is a device that passes a beam of monochromatic light through a solution, and measures the amount of light that was absorbed by this solution (Fig. 2.26). Using a prism, the device separates the light into beams with different wavelengths (as happens in a rainbow with the separation of the colors of white light). Thus, it can pass through the sample a beam of monochromatic light (of a single wavelength, or almost). The spectrophotometer detects what amount of light is absorbed at each wavelength.

The set of absorbances at various wavelengths for a compound is called an absorption spectrum and varies from substance to substance. If a substance is green, for example, then it lets pass or reflects the color at that wavelength, absorbing more light in the red region. Figure 2.27 presents spectra of various organic compounds.

Since different substances have different absorption patterns, spectrophotometry allows, for example, to identify substances based on their spectrum. It also allows quantifying them, since the amount of light absorbed is related to the concentration of the substance, according to **Beer-Lambert's Law**, seen previously.

Quite often it is necessary to quantify substances in complex mixtures, or that do not significantly absorb light at any wavelength. In these cases, the so-called colorimetric methods are used—the compound to be quantified is put in contact with a specific reagent, in order to develop a color whose intensity is directly proportional to the concentration of the substance in the original mixture.

The most common radiation sources are based on incandescence, but must operate at high temperatures to have appreciable coverage in the ultraviolet. They consist of

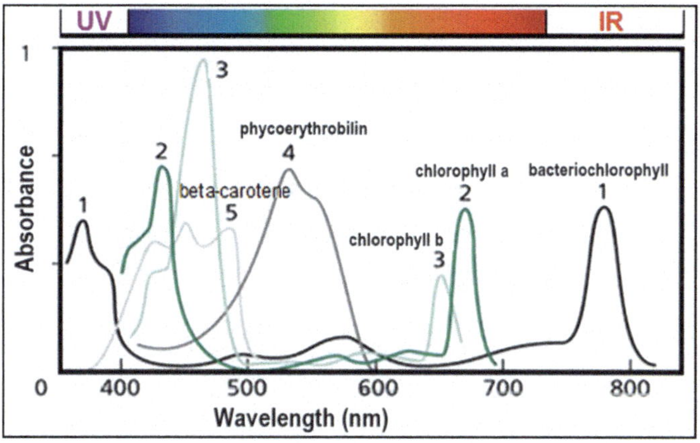

Fig. 2.27 Absorption spectra of different organic compounds

filaments of materials that are excited by electrical discharges with high voltage or electric heating.

For a source to be of good quality it must generate continuous radiation; have radiant power intensity sufficient to allow its detection by the detector system; be stable. In addition, it must have a long lifetime and low price. Examples of radiation sources are W lamps, quartz-iodine; hydrogen or deuterium discharge; hollow cathode lamp and LASER. Figure 2.28 compares the radiations provided by deuterium and tungsten lamps.

The monochromators have the function of selecting the wavelength in which one is interested in the analysis. They consist of an entrance slit of a radiation dispersion element, and exit slit. They can be of the reticular or prismatic type (Fig. 2.29).

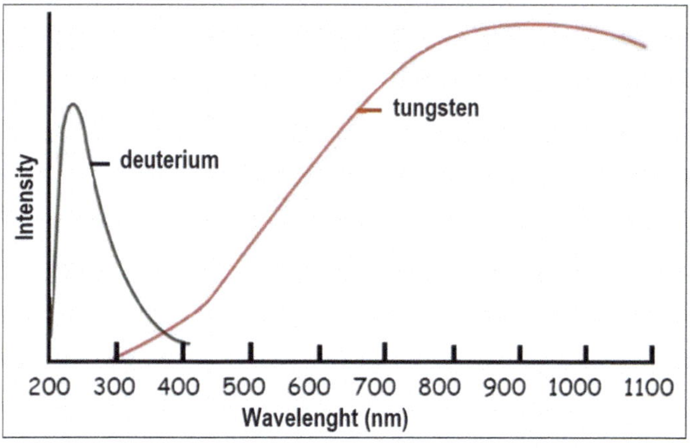

Fig. 2.28 Comparison between the radiations provided by the deuterium and tungsten lamps

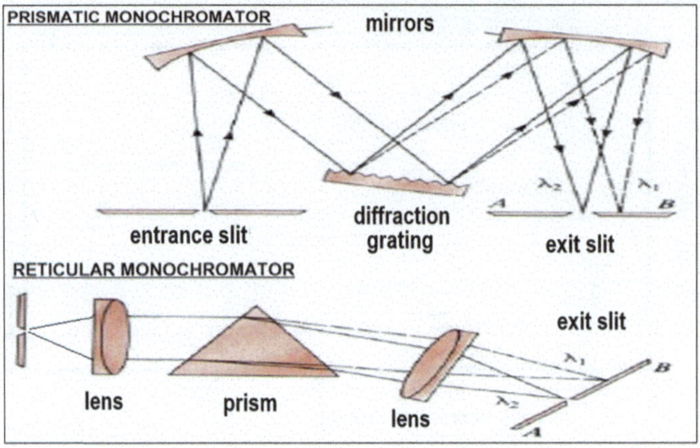

Fig. 2.29 Types of monochromators used in UV–VIS spectroscopy

With the reticular monochromator, the main dispersing element is the diffraction grating. This grating consists of a transparent or reflective plate with many parallel and equidistant grooves. The dispersion resulting from this grating is linear. The various dispersed wavelengths are equally spaced, so the exit slit will isolate a band of radiation of constant width. The resolution is much higher than in prisms.

With the prismatic monochromator, the polychromatic radiation coming from the radiation source passes through the entrance slit and hits the face of a prism, undergoing a deviation. The various wavelengths will have different directions after incidence on the prism. If a rigorously controlled adjustment of the exit slit is performed, the desired wavelength can be selected.

The spectrophotometers used in UV-VIS spectroscopy can be single or double beam (Fig. 2.30).

Single-beam ones are more laborious operationally, as the sample and the blank must be alternately placed in the single radiation beam. They are not suitable for measuring absorbances as a function of time.

In double-beam spectrophotometers, two radiation beams are formed in space, by a mirror that divides the beam coming from the monochromator into two. One beam passes through the reference solution (blank) to the transducer and another, at the same time, passes through the sample to the second transducer. The two currents will be determined and displayed on the signal indicator. With the help of an appropriate device, the difference in transmittance between the two beams is calculated; this difference will be displayed on the signal indicator as absorbance.

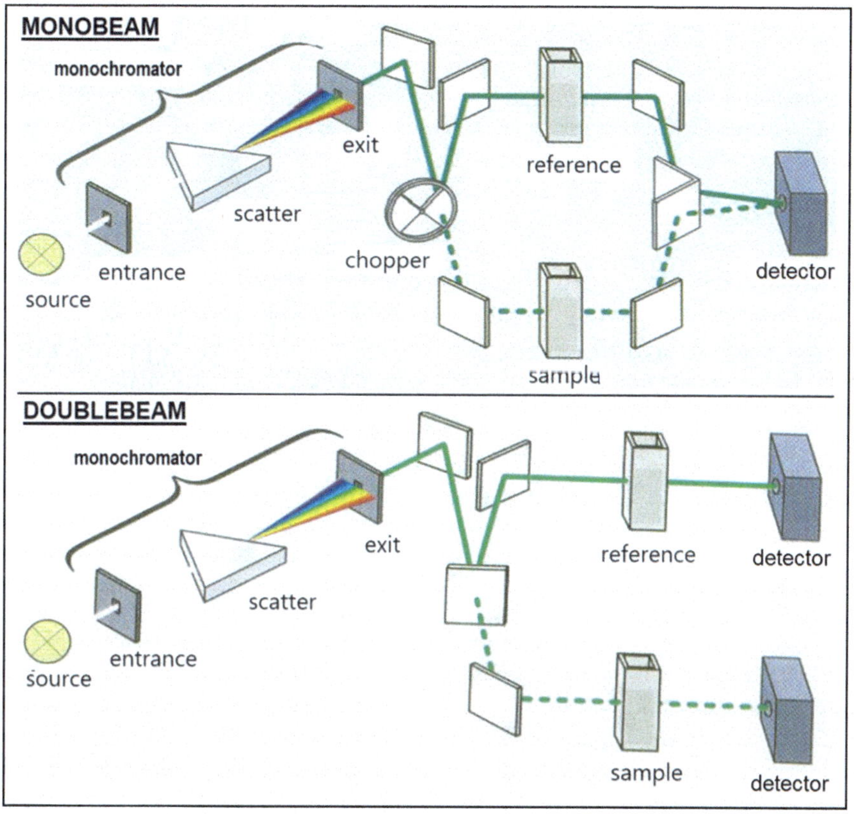

Fig. 2.30 Comparison between single and double-beam UV–VIS spectrophotometers

2.4.2 Applications of UV–VIS Spectroscopy

UV–VIS spectrometry can perform both qualitative and quantitative analyses. In qualitative analysis, depending on how much light the sample absorbs will determine what the species is, as the spectrum is characteristic of that particular chemical species.

In quantitative analysis, depending on the absorbance will determine the concentration of the sample. The special condition for any quantitative determination is the observation of **Beer-Lambert's Law**, but the control of pH, solvent extraction techniques, adjustment of the oxidation state, among others are also very important.

2.4.3 *Sample Preparation for UV–VIS Spectroscopy*

The preparation of the samples is similar to atomic absorption spectrometry. The samples must be in solution, or at least, the metals to be dosed must have been solubilized (see item 2.1.1.4).

2.4.4 *Sources of Error in UV–VIS Spectroscopy*

The sources of error in UV–VIS spectroscopy analysis are very similar to those of AAS, as the sample preparation and the principle of analysis are similar.

2.5 Inductively Coupled Plasma Atomic Emission Spectrometry (ICP)

ICP-AES (Inductively Coupled Plasma—Atomic Emission Spectrometry) is a technique of instrumental chemical analysis that makes use of a source of excitation from argon plasma at high temperature (7000–10,000 K) to produce, in a sample introduced in the form of mist in the center of the plasma, atoms; excited ions that emit radiation at wavelengths in the range of 125–950 nm (Fig. 2.31).

The emitted radiations, after suitable separation of their wavelengths by optical systems, have their respective intensities measured by means of *radiation detectors* specific (photo multipliers or solid state detectors) and correlated to the corresponding concentrations through calibration curves obtained by the prior measurement of Certified Reference Standards (CRM—Certificate Reference Material). This technique is also known as **ICP-OES** (Inductively Coupled Plasma Optical Emission Spectrometry) or, simply **ICP** (Fig. 2.32).

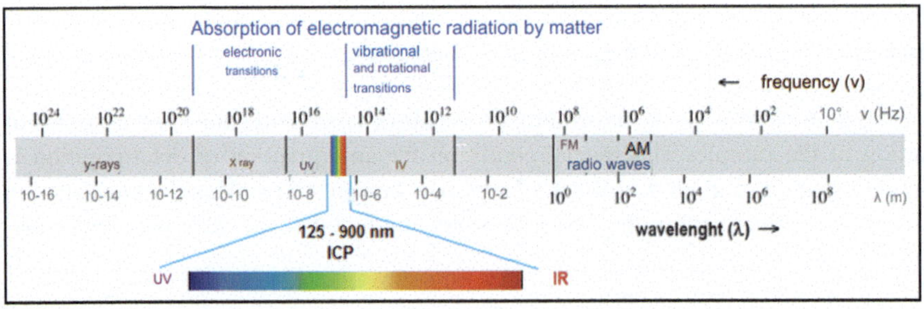

Fig. 2.31 Division of the regions of the electromagnetic spectrum, by wavelength and frequency. In detail, the range of wavelengths in the regions of ICP spectroscopy

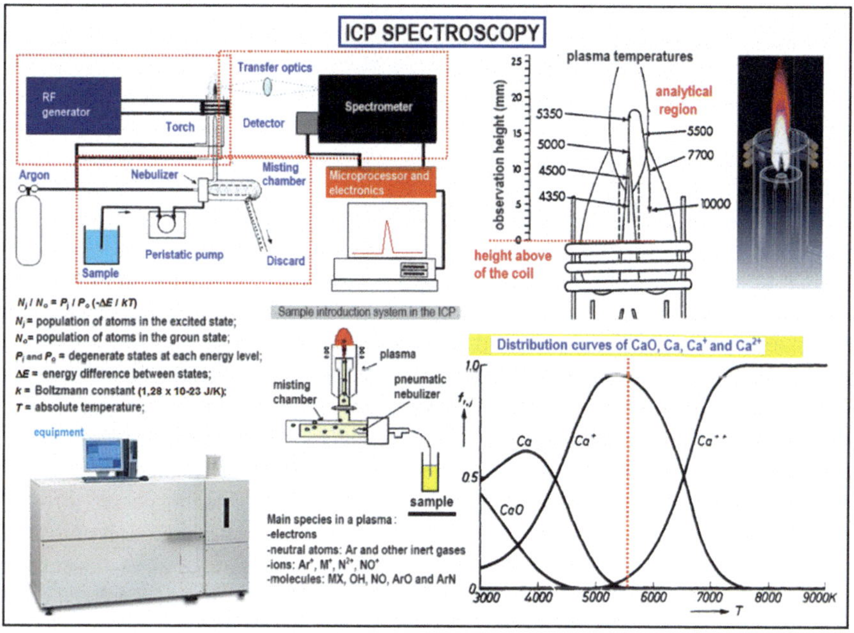

Fig. 2.32 Characteristics of the ICP spectroscopy analysis technique

The use of a plasma in energy generation allows increasing the population of excited species, according to Eq. 2.6, the Boltzmann Equation.

$$\frac{N_j}{N_o} = \frac{P_j}{P_o}.e^{-\left(\frac{\Delta E}{kT}\right)} \tag{2.6}$$

where:

N_j = population of atoms in the excited state;
N_o = population of atoms in the ground state;
P_j and P_o = degenerate states at each energy level;
ΔE = energy difference between states;
k = Boltzmann constant (1.28 × 10^{-23} J/K);
T = absolute temperature.

For typical temperatures of flames and electrothermal furnaces (2000–3000 K), the population of atoms in the ground state is on the order of 10^4 to 10^{15} times the population of atoms in the excited state. The higher the temperature, the larger the population of excited atoms (however, N_o will always be >>> than N_j).

Compared to thermal sources such as flames, furnaces, and electric discharges, plasma provides much higher temperatures. In addition, the use of plasma as atomization sources offers several advantages compared to atomic absorption in the determination of trace elements in raw materials: it is a technique for multiple elements and has a wide working range. Plasma sources offer a much easier method for handling gaseous and liquid samples. The spectrum of several elements can be recorded at the same time, which is very important when the sample is small. Plasma sources also allow the determination of non-metals such as chlorine, bromine, iodine, and sulfur.

2.5.1 Principle of ICP Spectroscopy

An ICP is basically composed of a vaporization and nebulization system (like an EAA), an optical system, a detection system for the response of the atom to excitation. The energy of the plasma is provided by a radio-frequency source (27 or 40 MHz). The energy content and the temperature of the ICP can be adjusted.

The plasma is formed by collision mechanisms between molecules and ions of argon in a magnetic field induced by radiofrequency. The torch that supports the plasma is formed by 3 concentric quartz tubes, surrounded by an induction coil through which energy is supplied (2 to 3 kW). The flow of argon passes through the torch and is ionized by the magnetic field produced by the induction coil. The magnetic field has axial force lines and the argon particles encounter resistance, producing heating and more ionization. The resulting ions and their associated electrons will interact with the fluctuating magnetic field. This generates enough energy to ionize argon atoms by shock excitation (Fig. 2.33).

The gas flow is seeded with free electrons that interact with the magnetic field, acquiring enough energy to further ionize the gas flow. A long and well-defined

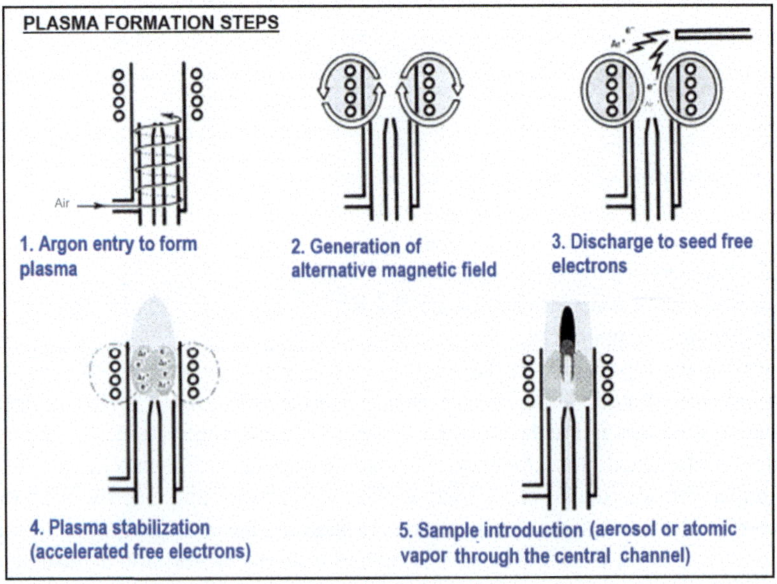

Fig. 2.33 Steps of plasma formation in ICP spectroscopy

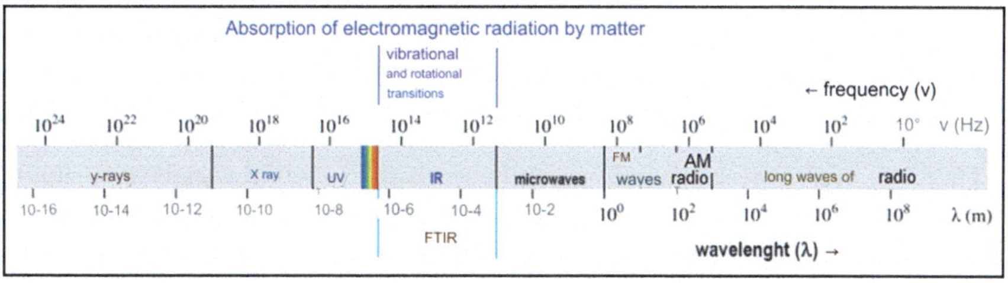

Fig. 2.34 Division of the regions of the electromagnetic spectrum, by wavelength and frequency. In detail, the range of wavelengths in the regions of IR spectroscopy (FTIR)

tail emerges from the top of the torch. This torch is the spectroscopic source, and contains all the atoms of the analyte and the ions that have been excited by the heat of the plasma, which is self-sustained by the continuity of the process. Argon flows through the three quartz tubes of the torch. The argon flow in the central tube carries the sample in aerosol form, from the nebulizer. The temperature obtained in the plasma, near the induction coil, is 10,000 K.

Argon is commonly used because it has high thermal conductivity to transfer heat from the outer region of the plasma to the central region. However, if the thermal conductivity is too high, it becomes difficult to maintain the plasma due to rapid cooling by the surrounding atmosphere. Argon also has suitable electrical resistivity for RF coupling and power dissipation, low spectral background (ICP OES) and is relatively low cost.

In addition to argon, helium and nitrogen can be used. It is important that the plasma gas is pure, as contaminants in the gas can extinguish the torch and/or interfere with the analysis results.

The sample in solution is taken to the plasma torch by a peristaltic pump, whose flow control can be regulated and should be kept constant during the calibration and analysis stages. A nebulizer converts the sample into aerosol which is introduced into the excitation area of the plasma. Figure 2.34 illustrates the conduction of the sample to the plasma via nebulizer. The analyzer can be a mono or polychromator. The detector is a photomultiplier that transforms the radiated energy into signals that can be measured.

2.5.2 *Applications of ICP Spectroscopy*

It is a technique especially indicated for extremely low concentrations in solution of a very large number of chemical elements (up to 70 in theory). The analysis of several elements can occur simultaneously, in a single sample, in a short period of time. The detection limits are excellent for chemical elements difficult to dose with AAS (As, Cd, Sn, W, Mo, Bi and In). Concentrations of these elements of the order of 1 to 20 ppm can be detected with HF resistant nebulizer, and 10 to 100 times lower with glass nebulizers (without hydride generator). For the other elements, it is possible to determine concentrations of the order of ppb in solution. The linearity is greater than in any other technique, covering, for example, concentrations from 10 ppm to 10^5 ppm, depending on the chemical element under analysis.

There are currently sequential or simultaneous ICPs, both for analysis of liquid and solid samples. In sequential equipment a monochromator moves to the point of the spectrum where the chosen line for dosage is located, scanning a small interval (0.1 nm) or fixing itself on the peak for quantification. In simultaneous ones, there are fixed channels scanning a wide range in the electromagnetic spectrum, allowing the determination of a large number of chemical elements in the same sample.

2.5.3 Sample Preparation for ICP Spectroscopy

The sample to be analyzed must be digested before the analysis, in order to dissolve the element(s) of interest. Therefore, they require the same preparation, with the same difficulties and interferences due to the matrix and the use of acids, mentioned for AAS (see item 2.1.1.4).

2.5.4 Sources of Error in ICP Spectroscopy

The emission spectra are complex interferences and inter-element are possible if the wavelength of the element of interest is very close to that of another element.

ICP is not suitable for the dosage of high contents due to its high sensitivity. Even when looking for less sensitive wavelengths and diluting the samples, the resulting oscillations will reflect errors due to instrumental variations, multiplied by the dilution factor, increased by errors due to the quality of the calibration of the glassware.

In ICP spectroscopy, due to the high temperature obtained in the plasma, the interferences of the type of formation of refractory oxides and deviations by ionizations, common in AAS, do not occur. However, the large number of lines generated by the extremely energetic source populates so intensely the electromagnetic spectrum that interferences by partial or total overlap of lines (*overlaps*) is very common. Salts in solution cause matrix differences that are reflected in the background, as in XRF.

ICP spectroscopy is quite suitable for routine analyses, provided that, as in XRF, a long instrumental calibration is carried out, first considering the spectral lines for each element, depending on the matrix to be investigated. Scans, for example, with 10 ppm of the element to be dosed and 1000 ppm of the element that constitutes the matrix should be overlapped for the choice of the wavelength free of spectral overlaps. At each marked change of composition of the samples (proportion of elements present) or at each different solubilization or dilution, the calibration should be redone.

2.6 Infrared Spectroscopy (IRS or FTIR)

Infrared spectrometry (IRS) uses absorption in the region of infrared of the electromagnetic spectrum for the chemical analysis of compounds. For this, a monochromatic infrared radiation is passed through the sample, and the amount of energy transmitted is recorded. Repeating this operation over a range of wavelengths of interest (normally 4000–400 cm^{-1}) a graph can be constructed, with "wave number" in cm^{-1} on the horizontal axis and transmittance in % on the vertical axis.

EIV works almost exclusively on covalent bonds, and is widely used, especially in Organic Chemistry. The high selectivity of the method makes it possible to identify a constituent of the sample of interest in a complex matrix. Figure 2.34 presents the division of the regions of the electromagnetic spectrum, by wavelength and frequency. In detail, the range of wavelengths in the regions of IR spectroscopy.

An infrared spectrometer is the instrument that produces and makes the infrared radiation pass through a sample and produces a spectrum where the absorption peaks are directed downwards on the transmittance axis. Because the chemical bonds in an organic molecule interact with infrared radiation, the IV spectrum offers a considerable amount of structural data. Figure 2.35 presents characteristics of IR spectrometry.

The **infrared spectroscopy by *Fourier transform*** (commonly referred to as FTIR—Fourier transform infrared spectroscopy) allows for a faster analysis, as it

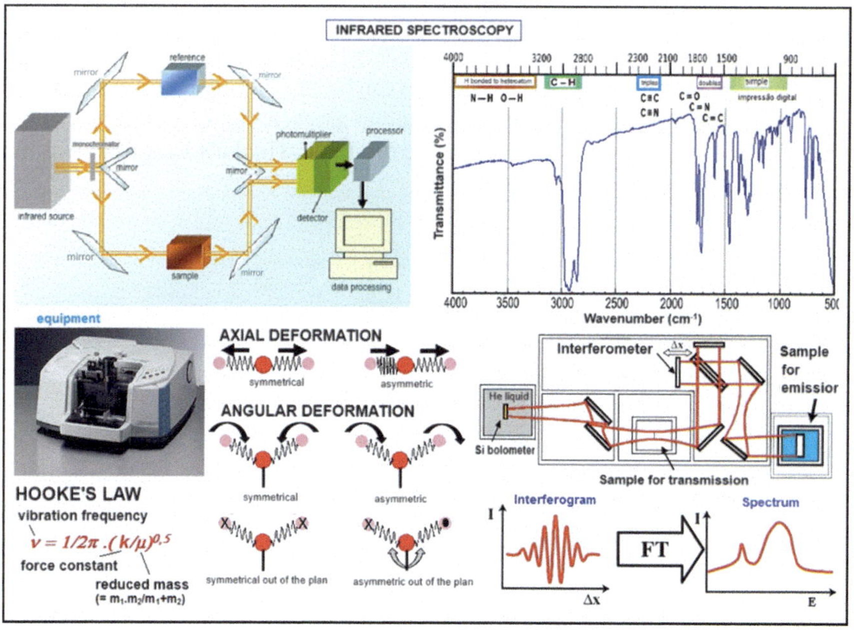

Fig. 2.35 Characteristics of the IV spectrometry analysis technique (FTIR)

conducts IR radiation through an interferometer, instead of collecting data by varying the frequency of monochromatic infrared light. The measured signal is the interferogram, which through a Fourier transform results in a spectrum identical to that of conventional (dispersive) IR spectrometry.

The spectrophotometers FTIR are cheaper than conventional ones because an interferometer is lower priced than a monochromator. In addition, the measurement of a single spectrum is much faster because information from all frequencies is collected simultaneously. This allows for multiple readings of the same sample and the average of them to be taken, thus increasing the sensitivity of the analysis. Due to its various advantages, modern infrared spectrophotometers are of FTIR type.

2.6.1 Principle of FTIR Spectroscopy

The principle of infrared spectroscopy is the measurement of the energy absorbed in the vibrational transitions of a sample. The movement of the atoms that make up the molecules results in rotations and molecular vibrations. Only vibrations are considered, as the energies involved in different forms of rotation are very close. The frequency of such vibrations depends on the geometry, the shape of the molecule's potential energy surface, the masses of the atoms, and possibly the vibronic coupling.

If the molecule receives electromagnetic radiation with exactly the same energy as one of these vibrations, then the radiation will be absorbed as long as certain conditions are met. The vibrational resonance frequencies are determined by the normal modes corresponding to the potential energy surface of the standard electronic state. Nevertheless, the resonance frequencies can be, in a first approximation, related to the bond length and the masses of the atoms at each end of it. Considering a molecule, in simplified terms, as a harmonic oscillator, **Hooke's Law** describes the frequency of vibration, according to Eq. 2.7.

$$\nu = \frac{1}{2}.\pi.\left(\frac{k}{\mu}\right)^{1/2} \tag{2.7}$$

where:

ν = vibration frequency in Hz.

k = force constant in Nm^{-1}

μ = reduced mass ($= m_1.m_2/m_1 + m_2$).

The solution to Eq. 2.7 indicates that average infrared radiation has enough energy to excite molecular vibrations to higher energy levels. The wavelength of the infrared absorption beams is typical of specific chemical bonds, such as those present in organic and organometallic molecules. However, the vibration will only be observed when it changes the dipole moment of the molecule. In practice, the application of electric fields with suitable frequencies (energies) causes the molecular vibrations to result in changes of dipole moment, due to the variation of the electronic distribution around the bonds. When the oscillating dipole is in phase with the incident electric field, energy transfer from the radiation to the molecule occurs, resulting in a transition.

In IR spectrometry, the **infrared source** consists of an inert solid that is electrically heated to temperatures between 1500 and 2200 K. Examples of radiation sources are Nernst emitter (rare earth oxides), Globar resistance and CO_2 laser.

The IR radiation from the source, collimated by a mirror, is introduced into a Michelson interferometer, which is a device formed by a beam splitter, a fixed mirror and a movable mirror (Fig. 2.36).

The radiation incident on the splitter is separated into two beams that are reflected again (one by the fixed mirror and the other by the movable mirror) towards the beam splitter. In this way, when these two parts are recombined, an interference process occurs. The result of this interference will depend on the difference between the optical paths traveled by each beam (which is determined by the distance of the movable and fixed mirrors to the beam splitter).

As the two beams combined by the splitter are in phase, the superposition of the maximum of one wave with the other results in a beam whose intensity is given by the sum of the intensities of each of the beams (Fig. 2.36a). On the other hand, the

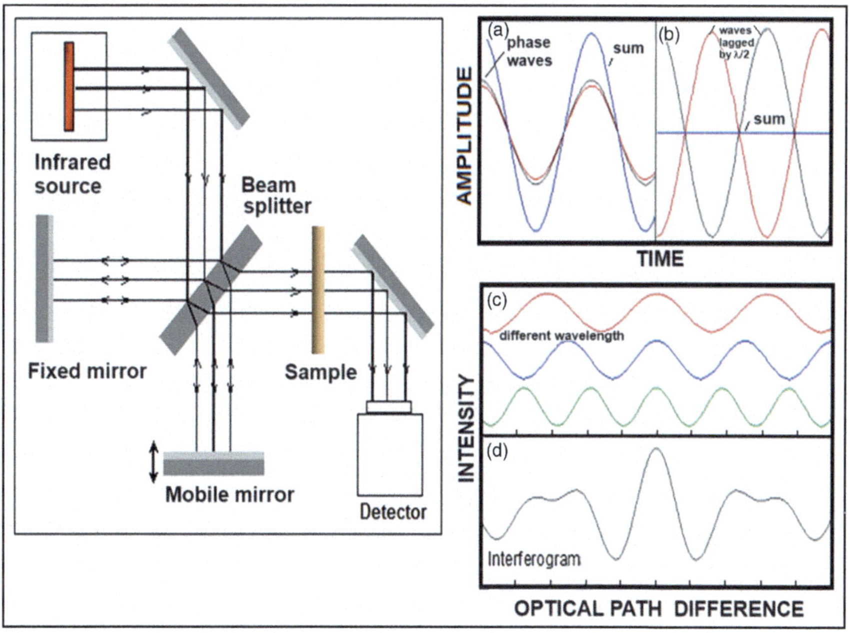

Fig. 2.36 Schematic representation of a measurement by Fourier transform infrared absorption spectroscopy. Sum of interferences (in blue): **a** two waves in phase; **b** out of phase waves by $\lambda/2$. Interference in waves with different wavelengths. **c** waves with different lengths. **d** Interferogram resulting from the superposition of the waves in (**c**)

path difference when the movable mirror has moved $\lambda/4$ will be $\lambda/2$. In this situation, the beams will be out of phase and the maximum of one wave meets the minimum of the other with equal amplitude, resulting in the cancellation of the radiation at that point (Fig. 2.36b).

The radiation that passes through the interferometer is directed to the sample and the light transmitted by the material is focused on a **detector**. Detectors can be classified into three categories: thermal detectors (short response time and lower sensitivity), pyroelectric (glycine trisulfide) and photoconductors (bolometers). The latter are the most sensitive detectors. They are based on the interactions between the photons and the semiconductor.

The detector consists of a thin film of a semiconductor material like lead sulfide, mercury cadmium telluride or indium antimonide deposited on a non-conductive glass surface and sealed to protect the semiconductor from the atmosphere. The lead sulfide detector is used for the near infrared region. For the middle and further regions, the mercury or cadmium telluride detector is used. It must be cooled with liquid nitrogen or helium to minimize changes.

From the detector, the **interferogram** is formed, which has on the horizontal axis the optical path difference (between the movable and fixed mirrors) and on the vertical axis the intensity of the radiation. The infrared spectrum (i.e., the curve of the transmitted intensity as a function of the wavelength) can be obtained from the Fourier transform of the interferogram (Fig. 2.35d).

2.6.2 Applications of FTIR Spectroscopy

Infrared spectrometry is widely used in both industry and scientific research, as it is a quick and reliable technique for dynamic analyses. FTIR has limited surface sensitivity (typical sampling volumes are ~0.8 mm), for a minimum analyte area of ~15 μm. It is capable of identifying organic functional groups and often specific organic compounds. The data that are obtained identify atomic groups and functional groups. Databases of characteristic spectra allow comparing the data with those of thousands of cataloged substances, enabling the identification of constituents in the sample. However, information about inorganic compounds is quite limited. It can be applied to volatile compounds as it does not require vacuum.

There are small devices that can be transported for field measurements. With the increasing technology in computational filtering and result manipulation, samples in solution can be measured accurately (water produces a broad absorbance band in the range of interest, which would give an unreadable spectrum without this computational treatment).

In the interpretation of the infrared spectrum, it is especially important to separate one from the other the regions of vibrations. Due to the constant multiple forces and the reduced masses, there are several areas within a spectrum, in which very specific vibrations should occur. If they are present, then one can already draw a conclusion

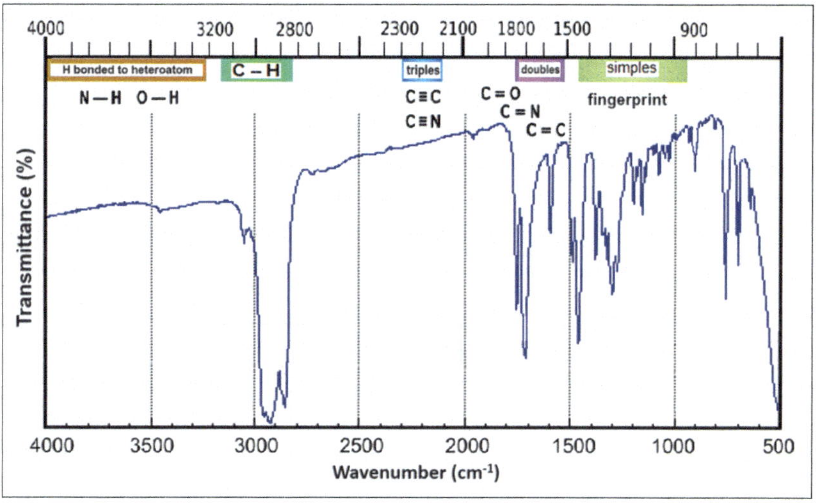

Fig. 2.37 Areas of the characteristic frequencies of different chemical bonds

about one of the elements present in the structure. Figure 2.37 shows the areas of the characteristic frequencies of different chemical bonds.

The area at 1500 cm^{-1} is considered the "fingerprint area". In this region, the specification of the vibration is very difficult. However, this area is the fingerprint of a bond, which means that it is recorded in the database. The identification of the compound is done by similarity.

In the area between 2500 and 3000 cm^{-1}, the **C–H** and **O–H** bonds can be identified. The vibration areas of 1600, 1700 and 1780 cm^{-1}, correspond to the presence of **C=O** and aromatic **C=C** bonds. At 1780 and 1700 cm^{-1}, there are the ester function and the carboxylic acid, respectively. The **Annex I** presents tables related to the analysis of bonds present in an IR spectrum.

The absorptions in this range do not apply only to bonds in organic molecules. IR spectrometry is also useful when it comes to the analysis of inorganic compounds (such as metallic complexes).

A variant of FTIR is the **near infrared spectrometry** (NIR). The components and the design of the NIR instrument are similar to UV–VIS absorption spectrometers. Near infrared radiation extends over a range of 800 nm–2500 μm (12,500–4000 cm^{-1}) and has enough energy to excite overtones and combinations of molecular vibrations at high energy levels. NIR spectroscopy is typically used in the quantitative measurement of organic functional groups, especially O–H, N–H, and C=O. Detection limits are typically 0.1% and the most common applications include pharmaceutical, agricultural, polymeric and clinical analyses.

In the analysis of mineral raw materials, for example, the identification of HO, CO, Si–O, SO and PO bonds, which absorb infrared radiation, is often of interest. In particular, FTIR is used to analyze H_2O, CO_2, CO_3^{-2}, SO_4^{-2} in minerals and glasses (inclusions), carbon-based materials, like coals and organic materials.

2.6.3 Sample Preparation in FTIR Spectroscopy

IR spectrometry can analyze gaseous, liquid or solid samples. Gaseous samples are inserted into a cell called cuvette, being ready for analysis.

Liquid samples are in the form they are found or in solution. The most common solvents are carbon tetrachloride (CCl_4) and carbon disulfide (CS_2). Chloroform, methylene chloride, acetonitrile and acetone are common solvents for polar materials.

Solid samples are typically prepared by mixing a certain amount of the sample with a highly purified salt that is transparent to IR radiation (usually potassium bromide or sodium chloride). This mixture is ground and pressed in order to form a pellet through which the radiation can pass.

2.6.4 Sources of Error in FTIR Spectroscopy

In addition to sample preparation, as with other spectrometric techniques, it should be prevented that fluctuations in the electrical energy of the source output affect the final results, since both the sample and the reference are affected in the same way. For the same reason, it should also be prevented the influence of variations in the result due to the fact that the source does not necessarily emit the same intensity of radiation for all wavelengths.

Although the IR spectrum is characteristic of the molecule as a whole, certain groups of atoms originate bands more or less at the same frequency, regardless of the structure of the molecule, making their identification difficult, or even, causing errors in the identification of compounds.

It should also be considered that the type of source and detectors and reflectors affect the range of wave numbers that can be accessed, as shown in Fig. 2.38, for globar sources with distinct detectors (MCT-A—cadmium and mercury telluride and TGS—triglycine sulfate) and reflectors (KBr and CaF_2).

2.7 Raman Spectrometry

The Raman effect is the inelastic scattering of light that falls on a molecule, with a shift in wavelength due to the interaction of the incident photons with the vibration of the molecules. The **Raman scattering**, primarily determined by the structure of the molecule, is extremely sensitive to the crystallographic structure, symmetry, and chemical composition of the sample. These factors define the levels of vibrational energy, which are characteristic of each molecule.

Raman vibrational spectroscopy is complementary to infrared, and the biggest difference is that Raman depends on changes in the polarization of molecules, while infrared spectroscopy depends on changes in their dipole moment. Figure 2.39

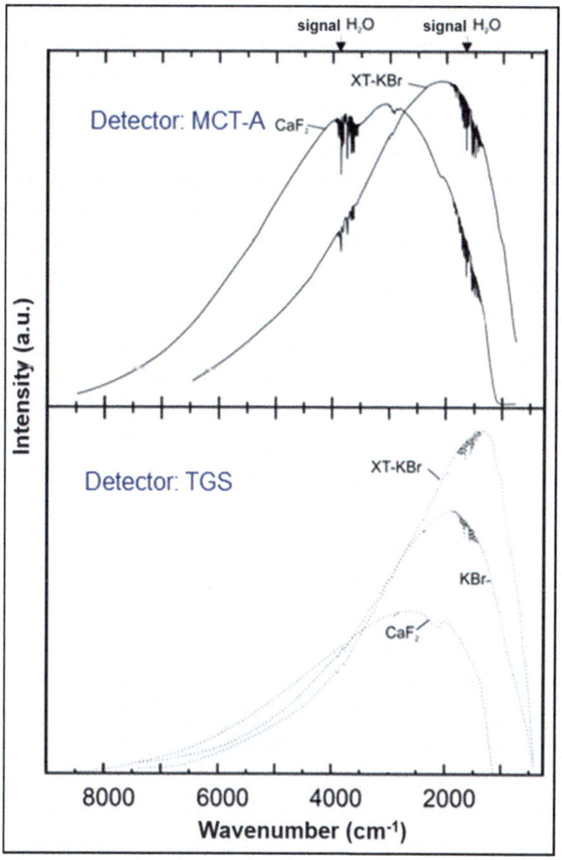

Fig. 2.38 Comparison in the IR spectrometer signal for globar sources with distinct detectors (MCT-A—cadmium and mercury telluride and TGS—triglycine sulfate) and reflectors (KBr and CaF_2)

presents the division of the regions of the electromagnetic spectrum, by wavelength and frequency. In detail, the range of wavelengths in the regions of Raman spectroscopy.

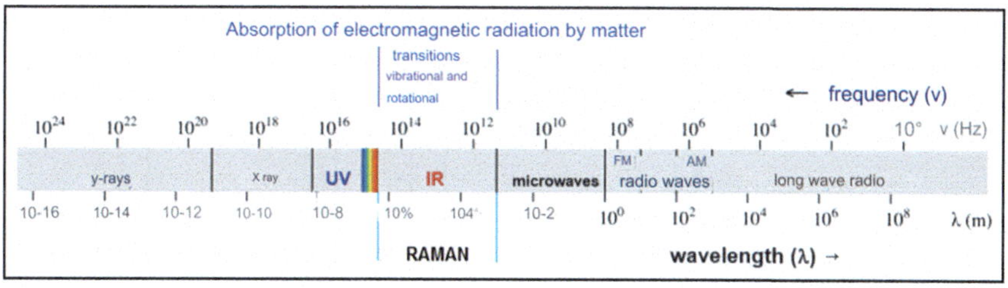

Fig. 2.39 Division of the regions of the electromagnetic spectrum, by wavelength and frequency. In detail, the range of wavelengths in the regions of Raman spectrometry

In Raman microscopy, Raman spectroscopy is combined with an optical microscope, and, thanks to the very reduced diameter of the laser beam used (which can be less than 1 μm) allows non-destructive analyses with high spatial resolution.

Raman spectroscopy uses a LASER to provoke in the sample the so-called "**Raman effect**", a change in the reflected radiation, outside the wavelengths of the visible. Raman spectroscopy allows the identification of minerals quickly and without preliminary preparation, as long as they can be placed under the objective lens of an optical microscope coupled to a Raman spectrometer.

2.7.1 *Principle of Raman Spectroscopy*

The Raman effect is an inelastic scattering process that can occur with a change in the vibrational, rotational or electronic energy of the molecule. It can be divided into 3 stages: absorption of the photon in an electronic excitation of the molecule; generation or absorption of a phonon and emission of a photon (Fig. 2.40).

Initially, the molecule is excited by LASER radiation to a virtual state and almost immediately undergoes relaxation, with a small energy difference corresponding to

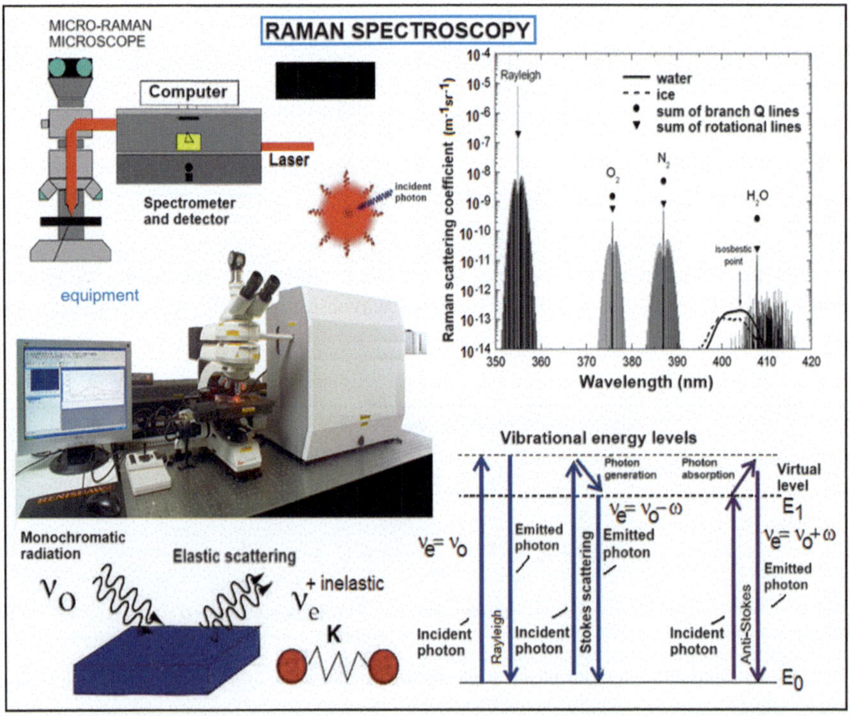

Fig. 2.40 Characteristics of the Raman spectrocopy analysis technique

the change in the vibrational energy of the molecule. The radiation that maintains the same frequency as the incident and does not reveal any information about the material is called Rayleigh scattering, but the one that changes, revealing the molecular composition, is known as Raman scattering. It can be of the Stokes and anti-Stokes type. In Stokes scattering, the incident photon is absorbed, a phonon is generated, and the excitation state relaxes (electron–hole pair recombine) by emitting a new photon. In anti-Stokes scattering, a thermally excited phonon is absorbed, and a photon with slightly higher energy than the incident photon is emitted.

The probability of Raman scattering occurring is very small (~1 in 10^7 photons) due to the fact that it involves excitation to a virtual energy level that is only allowed for a very short period of time. If the absorbed or emitted photon energy coincides with the energy of one of the molecule's electronic transitions, the probability of this process occurring becomes quite high. This phenomenon is referred to as resonant Raman spectroscopy.

The efficiency of Raman scattering increases when the energy of the laser excitation coincides with one of the optically allowed electronic transitions in the material. The intensity of Raman resonance depends on the density of electronic states available for optical transitions. When the effect is detected by ultra-sensitive filters and spectrometers, a vibrational spectrum characteristic of the sample is obtained, unique for each substance.

As the effect only occurs with a very small fraction of the incident photons, it is necessary to separate the original radiation from the altered fraction. For this, holographic filters (*notch*) have been developed that eliminate the laser light and only allow the passage of those few photons that have had their wavelength altered. This faint fraction is analyzed by a Czerny-Turner type optical spectrometer and the different wavelengths are simultaneously recorded on an ultra-sensitive CCD camera. A scheme of the arrangement of the different parts of a Raman spectrometer is in Fig. 2.41.

An important consideration in Raman spectroscopy is spectral resolution, the ability to resolve features within the spectrum. There are two ways to increase spectral resolution: by increasing the focal distance or by changing the grid used to disperse the spectrum. Doubling the focal distance approximately doubles the spectral resolution. Figure 2.42 illustrates this relationship.

2.7.2 Applications of Raman Spectrometry

For the identification of substances, it is necessary to compare the spectra obtained with standard spectra, from a library of known spectra.

The use of vibrational spectroscopy in the analysis of raw mineral materials is relatively recent. The conjunction of stable light sources and robust, self-aligned spectrometers with petrographic optical microscopes have enabled the technique to achieve micrometric spatial resolution, and thus a very powerful tool in mineralogy, since spectroscopic data can be spatially recorded with petrographic, chemical, and

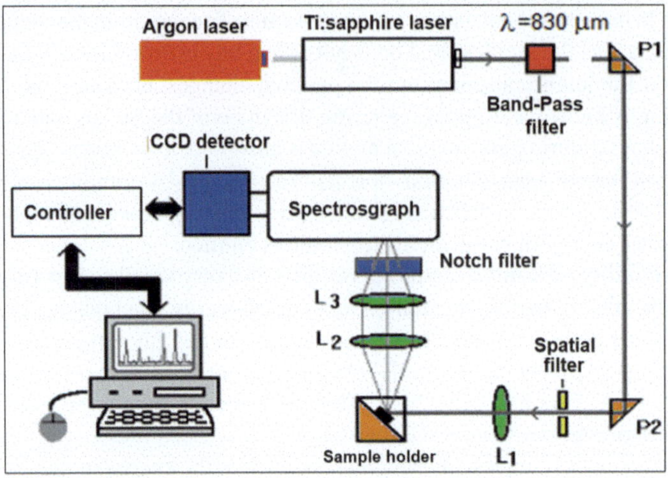

Fig. 2.41 Arrangement of the different parts of a Raman spectrometer

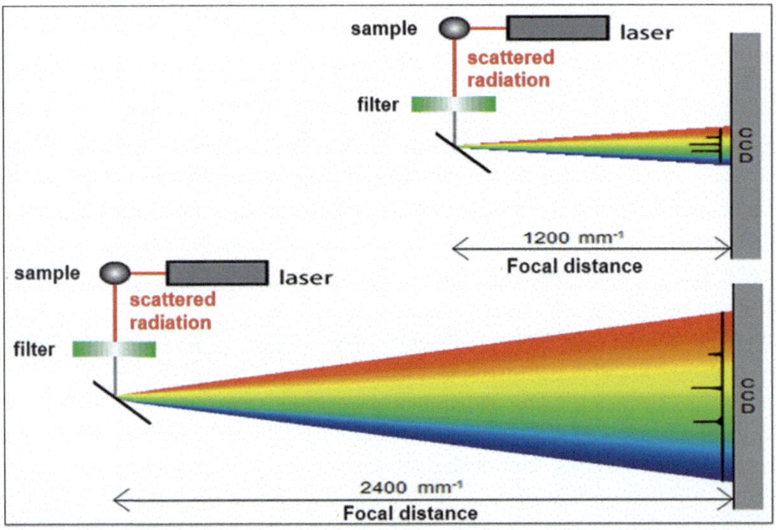

Fig. 2.42 Relationship between spectral resolution and focal distance in Raman spectrometry

diffractometric data. To illustrate, Fig. 2.43 presents spectrograms of some minerals such as fluorapatite, and distinct iron hydroxides and oxides.

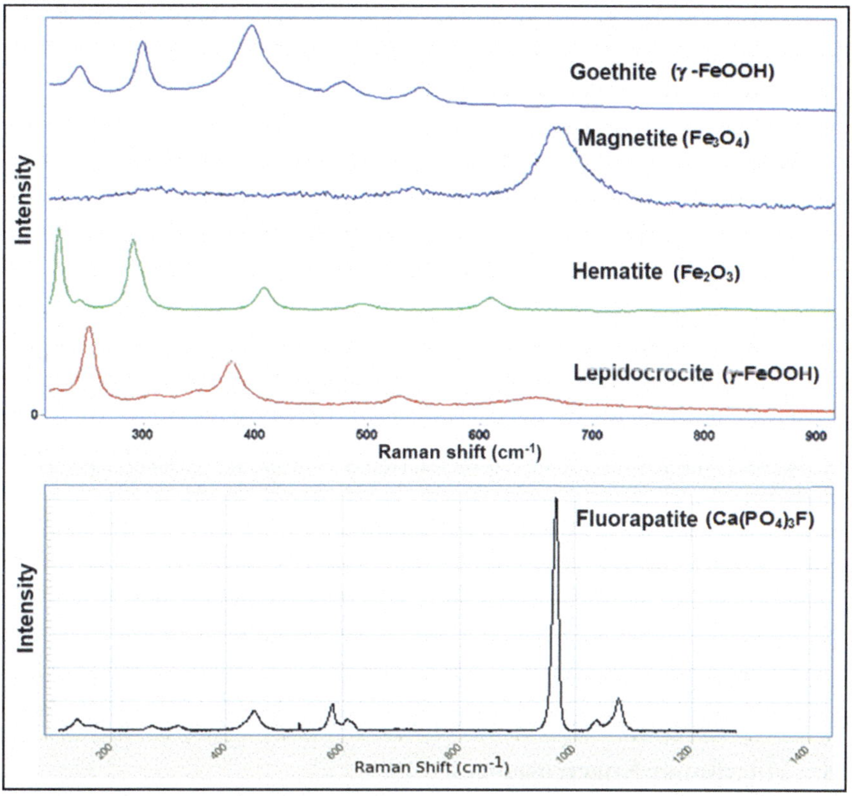

Fig. 2.43 Spectrograms of some minerals such as fluorapatite, and various iron hydroxides and oxides

2.7.3 Sample Preparation in Raman Spectroscopy

In terms of sample preparation, Raman spectroscopy has a series of advantages over other analysis techniques, as it can be used for the analysis of solid, liquid or gaseous samples. No sample preparation is necessary.

2.7.4 Sources of Error in Raman Spectroscopy

The main difficulty in Raman spectroscopy is to avoid overlap of the Raman signal by diffuse light from intense Rayleigh scattering. For this, Notch filters are used, which filter wavelengths within approximately 100 cm^{-1} of the laser wavelength.

Another aspect to consider is that the Raman effect is very weak, which gives it low sensitivity, making it difficult to measure low concentrations of a substance. This can be overcome by using one of the alternative techniques to intensify the Raman effect:

(i) Fourier Transform (FT) Raman: creates an interference pattern that can be analyzed to recover the spectrum. It has the advantage of being faster than the dispersion technique, but is limited in resolution and in the choice of laser wavelength;
(ii) Stimulated Raman Scattering: Raman signal 4 to 5 orders of magnitude larger. It is triggered by a strong LASER pulse, which excites a second mode, which excites a third and so on (cascade effect);
(iii) Resonance Raman Scattering (RR): the energy of the incident radiation is close to that of an electronic excitation energy. LASER*s* adjustable can be used to achieve this objective. The Raman scattering of the vibration modes around this excited state is then greatly enhanced by resonance effects.

Otherwise, the Raman effect can be interfered with by fluorescence from some materials.

Also, the choice of wavelength used is important, and can vary from near infrared to ultraviolet. As already mentioned, the choice may be limited by the density of the diffraction grid. Besides samples that show fluorescence it is vital to choose a wavelength that minimizes fluorescence, otherwise this will mask the weak Raman effect.

2.8 Mössbauer Spectroscopy

The **Mössbauer spectroscopy (MS)** is a spectroscopic technique based on the absorption, resonance and emission of γ rays by the atomic nucleus, therefore, it is considered a nuclear spectroscopy, very close to Nuclear Magnetic Resonance (NMR). This resonant emission and absorption were first observed by the German scientist Rudolf Mössbauer, hence named in his honor as **Mössbauer effect**.

Mössbauer spectroscopy applied to the study of mineral raw materials is based on the decay of ^{57}Co to ^{57}Fe by the release of a γ ray, along with other types of energy. MS is a very sensitive technique for investigations structural, chemical, and magnetic properties of raw materials. Figure 2.44 presents the division of the regions of the electromagnetic spectrum, by wavelength and frequency. In detail, the range of wavelengths in the Mössbauer spectrometry region.

The basic elements of a Mössbauer spectrometer are a source, sample, detector, and a unit to move the source or the absorber (Fig. 2.45). More commonly, this is done by moving towards the source and away from the sample, while the speed varies linearly with time.

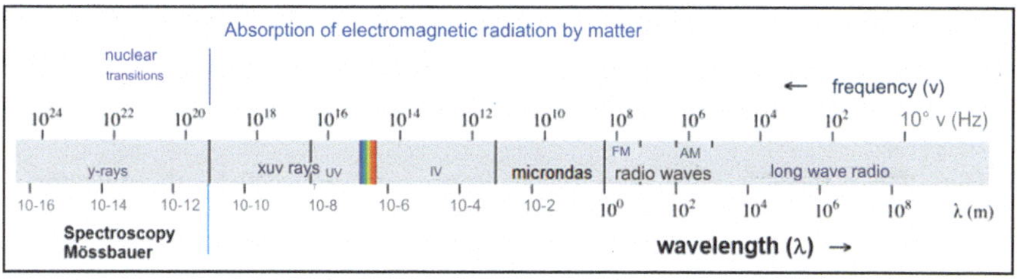

Fig. 2.44 Division of the regions of the electromagnetic spectrum, by wavelength and frequency. In detail, the range of wavelengths in the Mössbauer spectrometry region

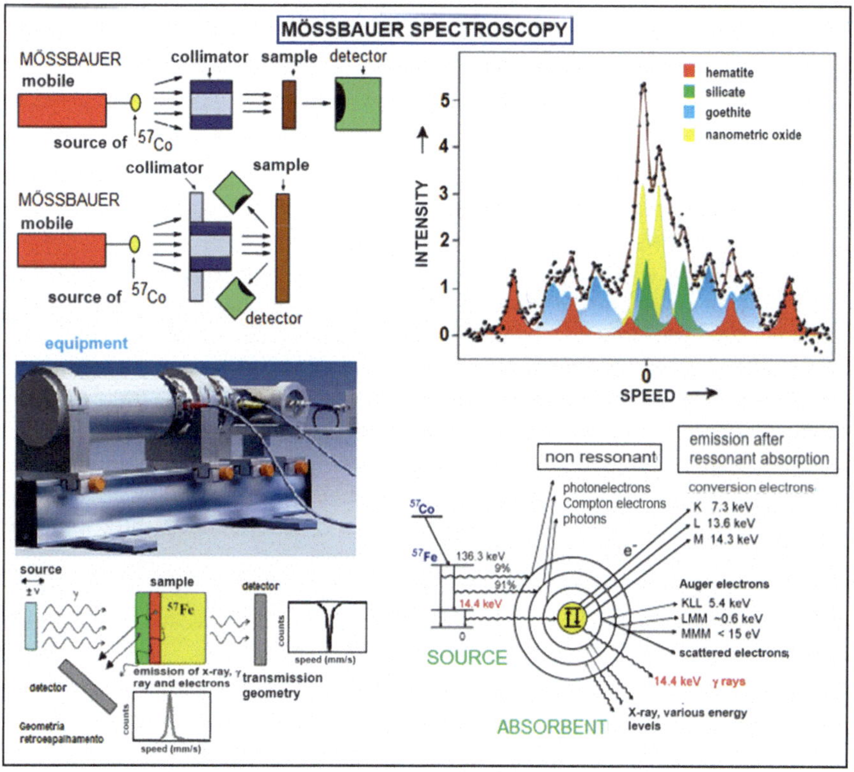

Fig. 2.45 Characteristics of the Mössbauer spectrometry analysis technique, including a Periodic Table indicating the active elements and those unsuitable for MS

2.8.1 *Principle of Mössbauer Spectroscopy*

Figure 2.45 presents the nuclear decay of ^{57}Co → ^{57}Fe and different backscattering processes for ^{57}Fe that can follow the resonant absorption of an incident γ photon. A photon that hits the sample containing the element with a nucleus identical to the emitter (in this case ^{57}Fe) can be absorbed or not, depending on the conditions of resonance. If not absorbed, the photon passes through the sample and hits the detector. If absorbed, the photon is re-emitted, but not necessarily in the direction of the detector.

Thus, when there is resonance, the photon count rate read by the detector is significantly lower than observed when the sample does not absorb the photons. This spectrometer geometry is known as **transmission geometry** and is the most used in obtaining the spectra.

The operation of the Mössbauer spectrometer is quite simple. The radioactive source (^{57}Co) is fixed at the end of a shaft connected to a transducer that performs an oscillatory longitudinal movement. The photon that is emitted by the source undergoes an energy modulation (E) by **Doppler effect,** given by Eq. 2.8.

$$E(V) = E(0)\left(1 + \frac{V}{c}\right) \tag{2.8}$$

where:

$V =$ instantaneous speed of the source;

$c =$ speed of light;

The emission of a photon in the Mössbauer effect, causes a recoil in the nucleus by the need for **conservation of momentum**, just like when a gun fires, there is a recoil in the opposite direction to the shot. If the nucleus is free (as in a gas) the recoil is not hindered during the emission or absorption of a γ ray.

However, if the nucleus is in a crystalline structure, not totally free for recoil, part of the recoil energy will be transformed into quantized vibrations of the crystalline network (phonons). However, if this energy is less than the necessary to reach the next higher energy level, the excitation of a mode vibrational has a low probability. The probability of no excitation occurring is given by the variable f (probability of recoil without causing vibration of the crystal lattice).

The Mössbauer effect occurs because in solids, the value of f is high enough for a recoil-free absorption to be possible. Thus, a ^{57}Co atom can decay to ^{57}Fe, which emits a γ ray, and can be absorbed without recoil by a neighboring ^{57}Fe. The emitting source is placed in oscillatory longitudinal motion so that the γ radiation is emitted with a modulated energy range through the **Doppler effect**.

This recoil-less emission process forms the basis of Mössbauer spectroscopy. Mössbauer spectroscopy detects and measures the γ rays emitted by a nucleus that are resonantly absorbed by a sample containing nuclei of the same isotope. In the resulting spectrum, the intensity of the γ rays is plotted as a function of source speed. At speeds corresponding to the resonance energy levels of the sample, a fraction of the γ rays is absorbed, resulting in a drop in intensity in the spectrum. Figure 2.46 presents the typical spectra of Mössbauer spectrometry. The number, positions, and intensities of the drops (absorbance peaks) can provide the following information about the sample:

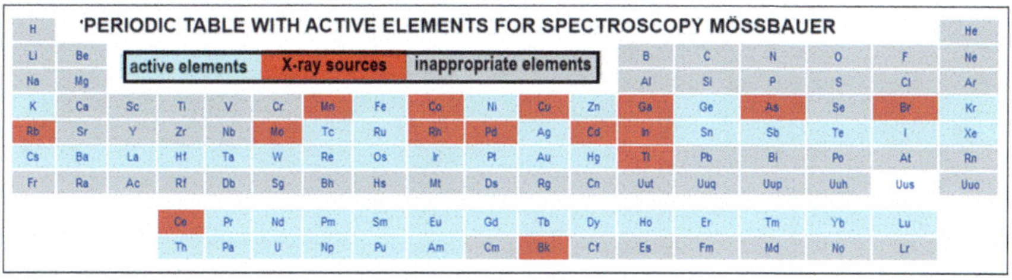

Fig. 2.46 Typical spectra of Mössbauer spectroscopy

(i) **oxidation state or valence** of the atoms that make up the sample, based on the **isomeric shift** (*IS*), defined by the single line. The Mössbauer effect appears in the transmission spectrum as a shift relative to zero speed; 1/2 and 2/3 represent the *spin* of the nucleus. It occurs in a non-magnetic material with cubic symmetry;

(ii) **local electrostatic potential** of the atoms, when a non-magnetic material with **quadrupolar splitting** (*QS*) different from zero. *QS* is defined by the interaction of the nuclear quadrupole moment with the electric field gradient, in relation to the energy levels of the nucleus;

(iii) **Magnetic moments** based on the **hyperfine field** (*HF*), when a cubic symmetry sample with a magnetic field interacts with the nucleus, producing a sextet pattern. In the simplest case, the areas of the lines vary in the proportion of 3:2:1:1:2:3.

Mössbauer parameters are sensitive to temperature, and this characteristic is sometimes exploited by using lower temperatures to improve peak resolution and induce magnetic phenomena of interest.

2.8.2 Applications of Mössbauer Spectroscopy

The Mössbauer spectrometry technique is widely used in the characterization of mineral raw materials. The interpretation of Mössbauer spectra of a mineral raw material, however, is often complex, involving: (i) number of resonance lines; (ii) number of subspectra; (iii) statistical dispersion of counting points; (iv) measurements as a function of temperature and (v) measurement under an applied magnetic field. The information obtained is often complementary to analyses by other methods, such as X-ray diffractometry, chemical analysis, thermal analysis, and microscopy (optical and electronic).

The identification and quantification of iron in minerals is conceptually straightforward, as the technique has specificity and reveals distinct oxidation states of Fe (Fe^0 (metal), Fe^{+2} and Fe^{+3}), as well as the type of coordination (tetrahedral, octahedral, etc.). Due to the high energy of the γ radiation and extremely narrow lines

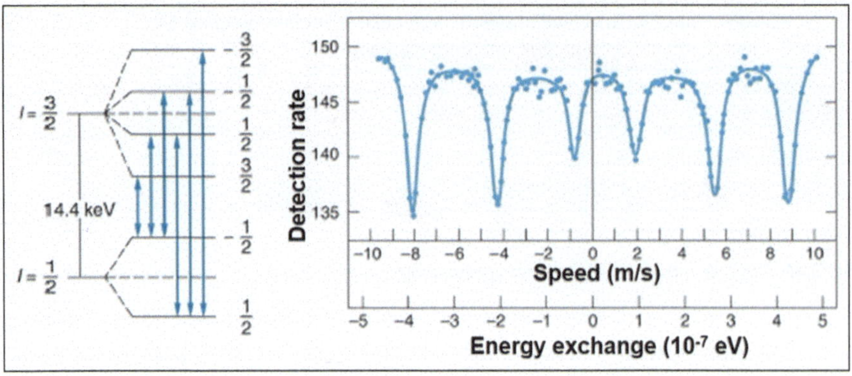

Fig. 2.47 Energy level diagram of the ^{57}Fe nucleus, showing the hyperfine fields of the excited state ($I = 3/2$) and the ground state ($I = 1/2$) of the Mössbauer spectrum of Fe_2O_3

in the spectrum, this technique is one of the most sensitive in terms of resolution, capable of detecting changes in just a few parts per 10^{12}.

Fe has low mobility and high abundance in the Earth's crust (fourth, after oxygen, silicon, and aluminum), which gives it characteristics of an indicator of geological processes. In addition, Fe represents more than forty times the sum of the abundances of all other magnetic elements in the crust. The magnetic properties of minerals are essentially derived from their iron content.

As an example, Fig. 2.47 presents the energy level diagram of the ^{57}Fe nucleus, showing the hyperfine fields of the excited state ($I = 3/2$) and the ground state ($I = 1/$ 2.) of the Mossbauer spectrum of Fe_2O_3, recording the absorption of γ rays emitted by ^{57}Fe by nuclei of ^{57}Fe in the sample as a function of the relative velocity of the source absorber.

2.8.3 *Sample Preparation in Mössbauer Spectrometry*

Sample preparation for Mössbauer spectroscopy is relatively simple. The sample to be analyzed is in powder form and scattered over a sample holder with a diameter equal to that of the detector window. The amount of sample used affects the resulting spectrum. If it is too small, some gamma rays γ will not reach the Fe atom. Using an excess sample can affect the area, intensity, width, and shape of the Mössbauer spectrum. If the chemical composition of the material is known, then the ideal amount of sample can be calculated. Generally, samples are mixed with an inert material, such as sucrose, to aid in the homogenization of the sample in the sample holder. The sample is fixed with a thin, non-absorbent material for gamma rays γ, such as cellophane.

2.8.4 Error Sources in Mössbauer Spectroscopy

Mössbauer spectra of mineral raw materials often exhibit a large overlap of peaks, resulting in difficulty in interpreting spectra. This is because the vast majority of minerals constituting rocks on Earth contain Fe^{+2} in octahedral coordination and thus have very similar Mössbauer parameters. For example, pyroxene, amphibole, and mica spectra are almost indistinguishable. For this reason, Mössbauer spectroscopy is not ideal for the identification of minerals (except for iron oxides, where magnetic properties can be extremely important).

Another fundamental issue is that Mössbauer spectroscopy works with solid samples, in which the representativeness of the whole sample is always a question to be addressed.

2.9 Nuclear Magnetic Resonance

Nuclear magnetic resonance (NMR) is based on the measurement of the absorption of radiofrequency (RF) radiation by a nucleus in a strong magnetic field. Absorption of radiation causes the nuclear spin to realign or oppose in relation to the direction of higher energy. After absorbing energy, the nuclei re-emit RF radiation and return to the state of lower energy. Figure 2.48 presents the division of the regions of the electromagnetic spectrum, by wavelength and frequency. In detail, the range of wavelengths in the NMR region.

Like other spectroscopic techniques, NMR depends on quantifiable energy variations that can be induced in small molecules when they are irradiated by electromagnetic radiation. The energy requirements of NMR (10^{-6} kJ/mol) are relatively small when compared to other spectrometric techniques (as seen, infrared radiation is in the 10^{-4} kJ/mol; ultraviolet in the range 160–1300 kJ/mol).

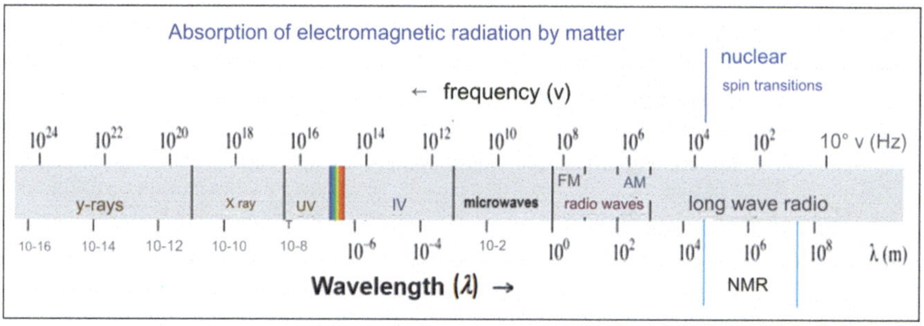

Fig. 2.48 Division of the regions of the electromagnetic spectrum, by wavelength and frequency. In detail, the range of wavelengths in the NMR region

In relation to Mössbauer spectroscopy, NMR has in common that both deal with nuclear resonance phenomena, but Mössbauer is not necessarily a magnetic resonance. For example, in Mössbauer spectroscopy the isomeric shift results from the fact that the nucleus has a defined size and the electron density at the nucleus is appreciable for s-level electrons. The chemical shift in NMR is a magnetic effect. Figure 2.49 presents some characteristics of NMR spectroscopy.

Nuclear magnetic resonance spectroscopy finds various applications in several scientific areas. In Chemistry, NMR spectroscopy is often used in the study of the

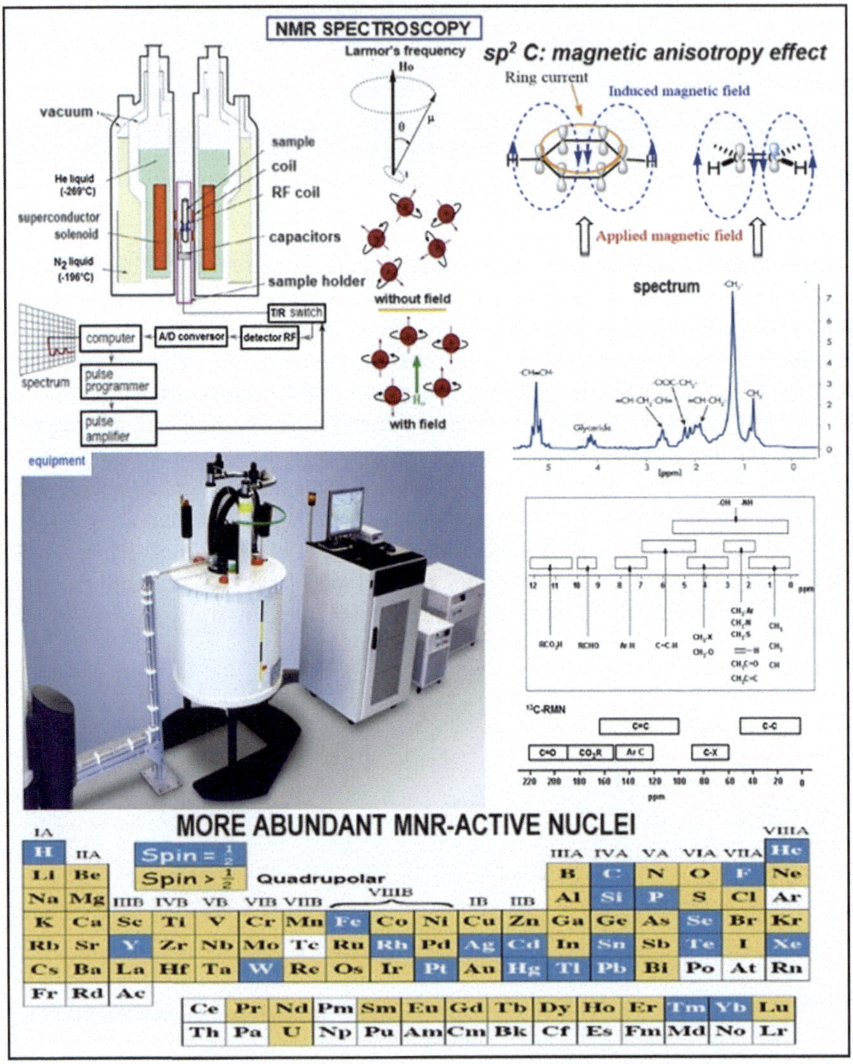

Fig. 2.49 Some characteristics of NMR spectroscopy

structure of compounds using simple one or two-dimensional techniques. It is a non-destructive technique that allows the analysis of organic compounds and some inorganic ones.

2.9.1 Principle of Nuclear Magnetic Resonance

Atomic nuclei are characterized by a rotation movement around an axis. The frequency with which the nuclei acquire precession movement is called **Larmor frequency**. Like all atomic movement, this is also quantized, in this case, by the quantum number of *spin* nuclear I, depending on the number of protons and neutrons, according to Table 2.3.

If I equals 0, the nucleus has no spin, therefore it will not produce a spectrum, since it has no magnetic properties. If I equals 1, the nuclei have what is called a quadrupolar electric moment, which means that their charge distribution is not spherical.

The principle of NMR is that nuclei with an odd number of protons, neutrons, or both will have an intrinsic nuclear spin (similar to the spin of electrons). When a nucleus with a spin different from zero is placed in a magnetic field, the nuclear spin can align in the same direction or in the opposite direction to the field. These two alignments of nuclear spin have different energies and the application of a magnetic field enhances this difference. A nucleus that has its spin aligned with the field will have a lower energy than when its spin is aligned in the opposite direction to the field.

The energy of an NMR transition depends on the strength of the magnetic field. The local chemical environment around a nucleus in a molecule will slightly disturb the magnetic field exerted on the nucleus, affecting its quantized transition energy. This dependence of transition energy on the position of a particular atom in a molecule makes NMR spectroscopy extremely useful for determining the structure of molecules and inorganic compounds.

When a compound containing hydrogen (therefore, protons) is placed under an external magnetic field, the protons can only assume one of two possible orientations in relation to the external magnetic field. The two possible alignments of the proton in a magnetic field do not have the same energy. When the proton is aligned in favor of the field (parallel-α) its energy is lower than the energy of when it is aligned

Table 2.3 Mass number, atomic number, and value of the nuclear *spin* for some chemical elements

Mass number (A)	Atomic number (Z)	Value of I	Examples
Odd	Odd	1/2, 3/2, 5/2, …	^{1}H, ^{19}F
Even	Even	0	^{12}C, ^{16}O
Even	Odd	1, 2, 3, …	^{14}N, ^{2}H=D
Odd	Even	1/2, 3/2, 5/2, …	^{13}C

against the magnetic field (antiparallel-β). Without a magnetic field, there is no energy difference between the protons, and the energy difference generated by the applied external field depends on the intensity of this field. It is then necessary a certain amount of energy to make the proton go from the lower energy state to the higher energy state, which in the ^{1}H-NMR spectrophotometer is provided by electromagnetic radiation (radiofrequency).

There are two types of NMR spectrometers: continuous wave (still commonly used in ^{1}H NMR spectroscopy for cost reasons) and, more recently, pulse or Fourier Transform (FT-NMR). In low-resolution continuous wave spectrometers, electromagnets are cooled with water, while in FT-NMR spectrometers, the magnets are cooled with liquid helium. Figure 2.50 schematically presents the continuous wave spectrometers and the FT-NMR.

A **continuous wave NMR** instrument consists of a magnet to separate the energy states of the nuclear *spin*, at least two radiofrequency channels, a field/frequency

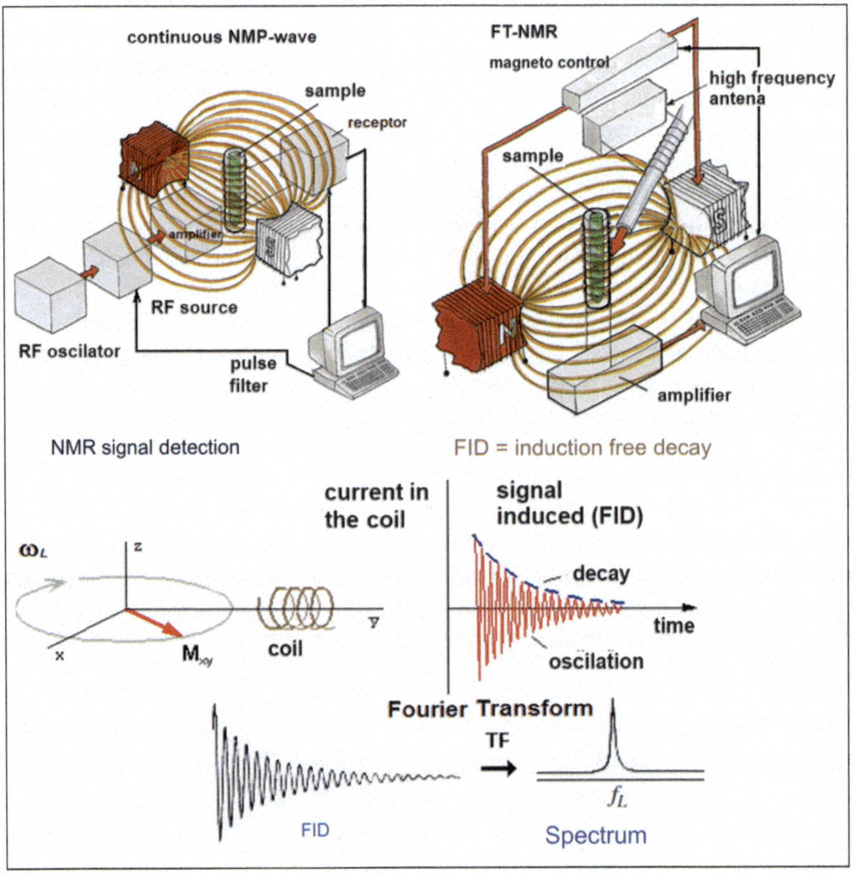

Fig. 2.50 schematically presents the continuous wave spectrometers and the FT-NMR

stabilizer, a device to provide RF energy irradiation, a sample holder containing coils for coupling the sample with the RF field, a detector to process the NMR signals; a generator to sweep both the magnetic field and the RF field, through the resonance frequencies of the sample, and a system of *display* of the resonance spectrum.

The spectrum is scanned by field scanning or frequency scanning. In frequency scanning, the magnetic field is kept constant, which keeps the energy levels of the nuclear spin constant. The RF signal is then scanned to determine the frequencies at which energy is absorbed. In the field scanning method, the RF signal is kept constant, the magnetic field is scanned, which causes variations in energy levels, which is then used to determine the magnetic field intensity that produces resonance at a given resonance frequency.

Fourier transform spectrometers use an RF radiation pulse to cause the align ment of the nuclei in a magnetic field in the direction of higher energy. The duration of the RF pulse is 1^{-16} s, enough to excite the nuclei, simultaneously. The interval between the pulses T can be from one to several seconds. During T, an RF signal in the time domain, called free decay (FID), is emitted when the nuclei return to their original state.

Free decays (FID) can be detected with a radio-receiver coil, perpendicular to the static magnetic field. The FID signal is digitized and processed on a computer. Common decay signals in the time domain of a pulse sequence can be summed, and an average obtained to improve the signal-to-noise ratio. The result is then converted into a signal in the frequency domain by a Fourier transform.

2.9.1.1 Analysis Parameters in NMR

The NMR spectrum is a graph of the **applied radio frequency** against the **absorption**. A signal in the spectrum is referred to as a **resonance**. In a molecule, some hydrogen nuclei are in regions of higher electron density than others, so there are protons that absorb energy in magnetic fields of slightly different intensities. As a result, signals appear in the NMR spectrum at different positions, called **chemical shift** (δ).

The absolute chemical shift is defined by the resonance frequency, expressed with reference to a standard compound, defined as being 0 ppm. Frequency changes are extremely small compared to the fundamental NMR frequency. The typical frequency change can be 100 Hz, compared to a fundamental NMR frequency of 100 MHz, so the chemical shift is usually expressed in parts per million (ppm), according to Eq. 2.9.

$$\delta = \frac{\left(\nu_{signal} - \nu_{reference}\right)}{\nu_{spectrometer}} .10^{6} \tag{2.9}$$

where:

ν_{signal} = signal frequency

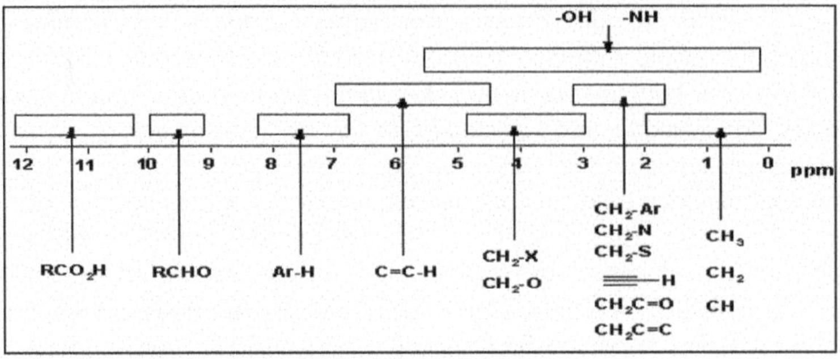

Fig. 2.51 Chemical shifts of different organic compounds

$\nu_{\text{reference}}$ = reference frequency
$\nu_{\text{spectrometer}}$ = spectrometer frequency.

Just as the number of signals in an NMR spectrum informs how many types of protons exist in a molecule, the position of the signals assists in determining what type of protons they are: aromatic or aliphatic; primary, secondary, or tertiary; benzylic, vinylic or acetylenic; adjacent to halogen atoms or to any other group of atoms. Figure 2.51 presents the chemical shifts of different organic compounds, typical of protons to be influenced by a single group. In cases where a proton is influenced by more than one group, the effects are essentially cumulative.

These different species of protons have different electronic environments that thus determine the exact location of the spectrum where the proton absorbs. This is what makes NMR so useful for the determination of structure, otherwise, all protons would have the same chemical shift. In this way, if the induced field opposes the applied field, the proton is **shielded**. However, if the induced field strengthens the applied field, then the field sensed by the proton undergoes an increase and the proton is said to be **deshielded** or unprotected. The influencing factors can thus be described:

(i) **Electronegativity**: The electrons around the proton create a magnetic field that **opposes** the applied field. As the field experienced by the proton defines the difference in energy between the two states of *spin*, the frequency and, therefore, the chemical shift δ/ppm will change, depending on the electron density around the proton. Electronegative groups associated with C-H bonds decrease the electron density around the protons, and there is less shielding, causing the chemical shift to increase. Table 2.4 exemplifies the effect of the electronegativity of atoms on the chemical shift of compounds.

These effects are cumulative, therefore, the presence of more electronegative groups further decreases the shielding and, therefore, greater chemical shifts. These inductive effects are not only felt by the immediately adjacent protons, but throughout the chain (Table 2.5). However, the effect weakens quickly as you move away from the electronegative group, as indicated by the chemical shifts of a bromide linked to –CH groups (Table 2.5).

Table 2.4 Effect of the electronegativity of atoms on the chemical shift of CH_3X compounds

Compound, CH_3X	CH_3F	CH_3OH	CH_3Cl	CH_3Br	CH_3I	CH_4	$(CH_3)_4Si$
X	F	O	Cl	Br	I	H	Si
Electronegativity of X	4.0	3.5	3.1	2.8	2.5	2.1	1.8
δ/ppm	4.26	3.4	3.05	2.68	2.16	0.23	0

Table 2.5 Reduction of the effect of the electronegativity of atoms on chemical shift as a function of the distance of the involved nuclei

Compound	CH_4	CH_3Cl	CH_2Cl_2	$CHCl_3$
δ/ppm	0.23	3.05	5.30	7.27
H signal	$-CH_2-CH_2-CH_2Br$			
δ/ppm	1.25 1.69 3.30			

(ii) **Magnetic anisotropy**: Magnetic anisotropy means that there is a non-uniform magnetic field. Electrons in π bonds (for example, aromatic compounds, alkenes, alkynes, carbonyls, among others) interact with the applied field which induces a magnetic field, causing a magnetic anisotropy. As a result, the nearby protons will experience three fields: the applied field, the shielding field of the valence electrons, and the field due to π bonds. Depending on their position in this third field, the proton can be shielded (lower δ) or deshielded (higher δ), which implies that the energy, and the absorption frequency, will be modified.

Aromatic protons of benzene rings are not shielded, and absorb energy in a low-intensity magnetic field. On the other hand, protons attached to carbons of double and triple bonds have relatively high shielding, due to the high electronic density of the π bonds, and absorb energy in a higher magnetic field. Figure 2.52 presents these interactions related to the shielding and deshielding of protons.

(iii) **hydrogen bonding**: protons that are involved in hydrogen bonds, such as in –OH or –NH, have a wide range of chemical shift values. The more the bond has hydrogen, the more the proton is deshielded and the greater will be the chemical shift. However, it can often be difficult to predict this influence because the amount of hydrogen bonds present is susceptible to factors such as hydration, acidity, concentration, and temperature.

The reference point from which chemical shifts are measured is not, for practical reasons, the signal of the nuclear proton, but the signal of a certain compound, usually, the TMS, tetramethylsilane $(CH_3)_4$ Si. TMS contains 12 protons, but these are all chemically equivalent, so they originate a single very clear signal. As silicon is less electronegative than carbon, the protons of TMS are in regions of high electronic density. Therefore, they are very shielded and the signal occurs in a region of the spectrum where few hydrogen atoms absorb energy, so this reference compound

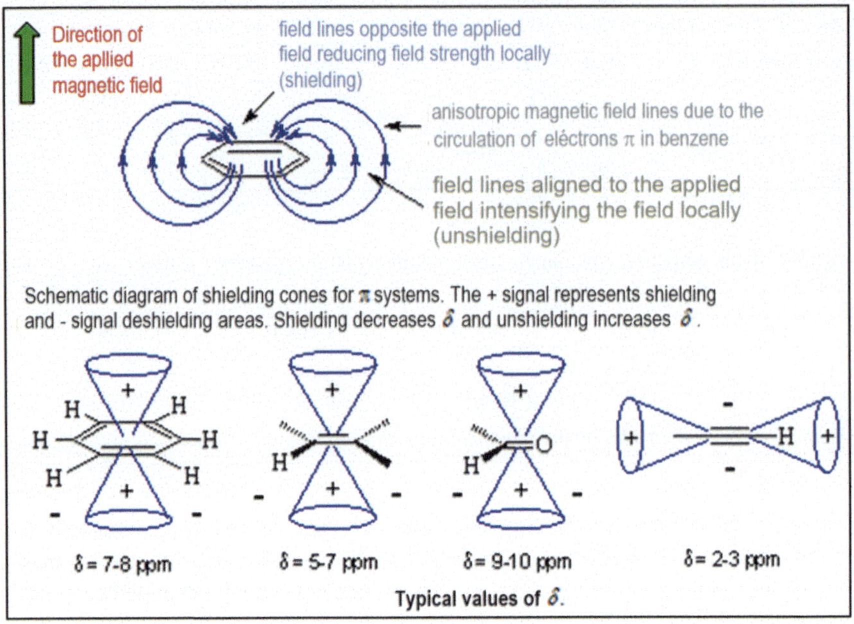

Fig. 2.52 Interactions related to the shielding and deshielding of protons

rarely interferes. with the signals from the other protons. After the spectrum has been raised, TMS can be easily removed by evaporation.

Coupling in ¹H-NMR. Spectra generally have peaks that appear as groups of peaks due to **coupling** with neighboring protons. Figure 2.53 presents the spectrum of 1,1-dichloroethane. The peak in the spectrum for δ equal to 2.1 ppm, for the integration 3H, is associated with the group $-CH_3$, as well as the one related to δ equal to 5.9 ppm, for unshielded 1H integration, is associated with $-CHCl_2$.

Coupling arises due to the magnetic field of adjacent protons that influences the magnetic field that the protons experience. Considering the effect of –CH on adjacent $-CH_3$, –CH can adopt two alignments with respect to the applied field. As a result, the signal for the adjacent $-CH_3$ is split into two lines, of equal intensity, a **doublet**.

Now considering the effect of $-CH_3$ on the adjacent –CH, the protons of $-CH_3$ give rise to eight possible combinations in relation to the applied field. However, some combinations are equivalent and there are four magnetically different effects. As a result, the signal for the adjacent -CH is split into four lines, with an intensity ratio of 1:3:3:1, a **quartet**.

The proximity of n equivalent H in neighboring carbon atoms causes the signals to be separated into $n + 1$ lines. This is known as the **multiplicity** or **coupling pattern** of each signal. Equivalent protons (or with the same chemical shift) do not couple with each other. If the neighbors are not all equivalent, more complex patterns arise. For a first approximation, protons in C sp^3 tend to behave as if they were equivalent.

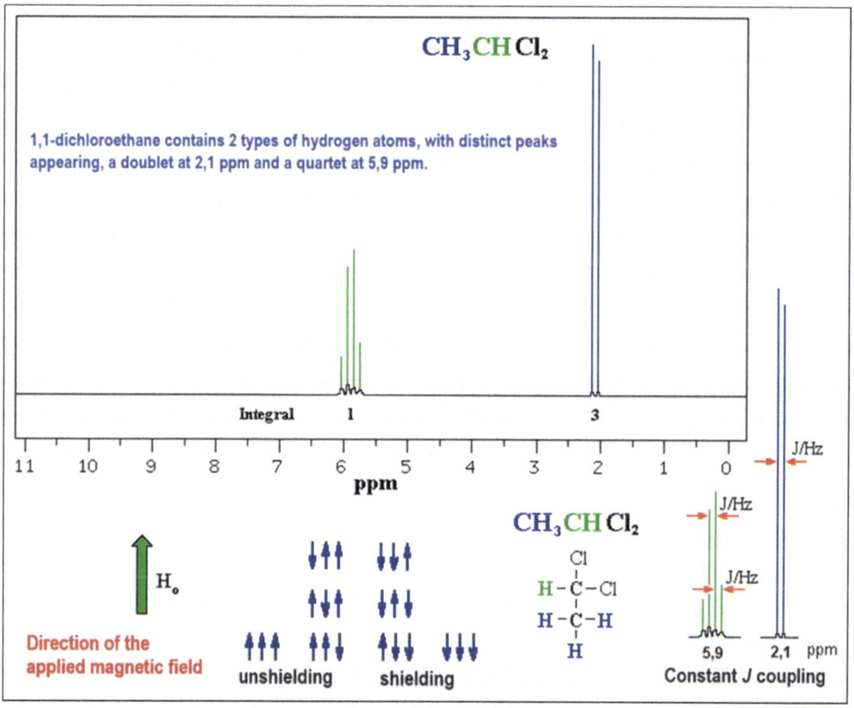

Fig. 2.53 NMR spectrum of 1,1-dichloroethane, highlighting the coupling between adjacent protons and its effect on shielding. In detail, the constant coupling J in the same compound

For the application example of the rule "$n + 1$" for 1,1-dichloroethane, it is found that for the δ of 5.9 ppm, a quartet, in unshielded 1H integration, is consistent with the $-CHCl_2$ next to a $-CH_3$ unit ($n = 3$, then $n + 1 = 4$ lines). And for the δ of 2.1 ppm, a doublet, in 3H integration, is consistent with the $-CH_3$ unit, to side of a –CH– ($n = 1$, then $n + 1 = 2$ lines).

Constant coupling *J*. The coupling constant J (usually in frequency units, Hz) is a measure of the interaction between a pair of protons. In a general type system, H_a –C–C–H–$_b$, a coupling with neighboring protons generically H_a with H_b, represented by J_{ab}, should be equal to the coupling of H_b with H_a, J_{ba}, therefore, $J_{ab} = J_{ba}$. The implications are that the spacing between the lines in the coupling patterns is the same, as seen in detail in Fig. 2.52, in the coupling patterns of H-NMR spectra of 1,1-dichloroethane.

The relationship of the intensity lines in a bonding pattern is given by a binomial expansion predicted by Pascal's triangle, as per Table 2.6. Thus, for H-NMR a proton with zero neighbors, n = 0, appears as a single line, a proton with one neighbor, $n =$ 1 as two lines of equal intensity, a proton with two neighbors, $n = 2$, as three lines of intensities 1: 2: 1, etc.

Table 2.6 Multiplicity of couplings according to the binomial expansion predicted by **Pascal's triangle**

Multiplicity	Intensity ratio
Singlet (*s*) singlet (*s*)	1 1
Doublet (*d*) Doublet (*d*)	1:1
Triplet (*t*) triplet (*t*)	1:2:1
Quartet (*q*) quartet (*q*)	1:3:3:1
Quintet Quintet	1:4:6:4:1
Sextet Sextet	1:5:10:10:5:1
Septet Septet	1:6:15:20:15:6

Coupling combined with chemical shift, (and proton integration) informs not only about the chemical environment of the nuclei, but also the number of neighboring nuclei active for NMR within the molecule. In more complex spectra with multiple peaks for similar chemical shifts or in spectra of other hydrogen nuclei, coupling is often the only way to distinguish different nuclei.

Second-order coupling (or strong coupling). The description above assumes that the coupling constant is small compared to the difference in NMR frequencies between non-equivalent spins. If the chemical shift separation decreases (or the coupling strength increases), the multiplicity patterns are first distorted, and then become more complex and less easily analyzed (especially if more than two spins are involved). It is possible to obtain the intensification of some peaks in a multiplet at the expense of the others, which sometimes almost disappear in the background noise, although the integrated area under the peaks remains constant. In most high-field NMR, however, distortions are generally small and characteristic distortions can in fact help identify peaks of interest. Figure 2.54 presents the spectrum of isopropanol containing multiplicity of couplings.

2.9.1.2 C-NMR Spectroscopy

A comparison between ^{1} H-NMR and ^{13}C-NMR allows us to observe the following differences and similarities:

(i) ^{13}C has only about 1.1% natural abundance (of carbon atoms);
(ii) ^{12}C does not have NMR behavior;
(iii) the nucleus of ^{13}C is about 400 times less sensitive than H nucleus to NMR phenomena;
(iv) due to low abundance, there is usually no coupling ^{13}C– ^{13}C;
(v) the chemical shift is usually between 0 and 220 ppm (Fig. 2.55);
(vi) similar factors affect the chemical shift in ^{13}C as seen for ^{1}H-NMR;
(vii) very long relaxation times (excited state to ground state) means absence of integrations;
(viii) "normal" ^{13}C spectra are of broad band protons;
(ix) number of peaks indicates the number of types of C.

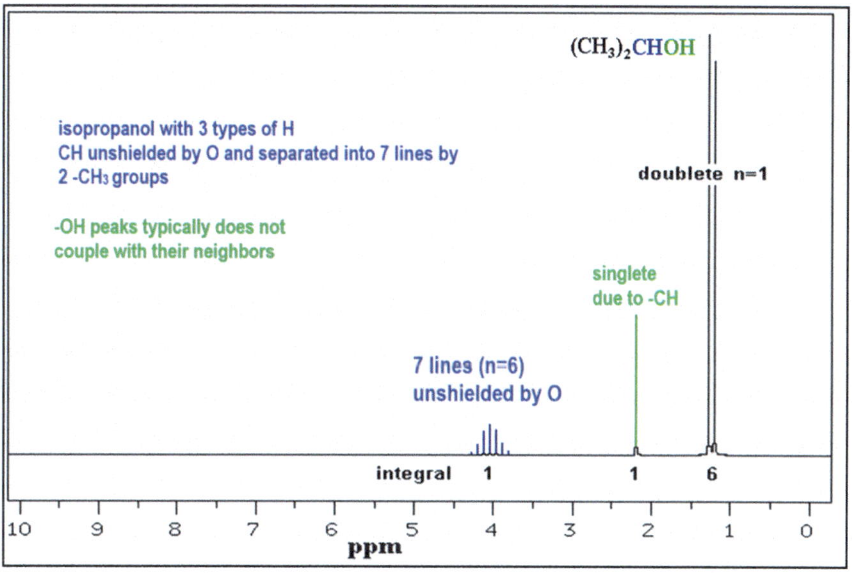

Fig. 2.54 NMR spectrum of isopropanol containing a multiplicity of coupling

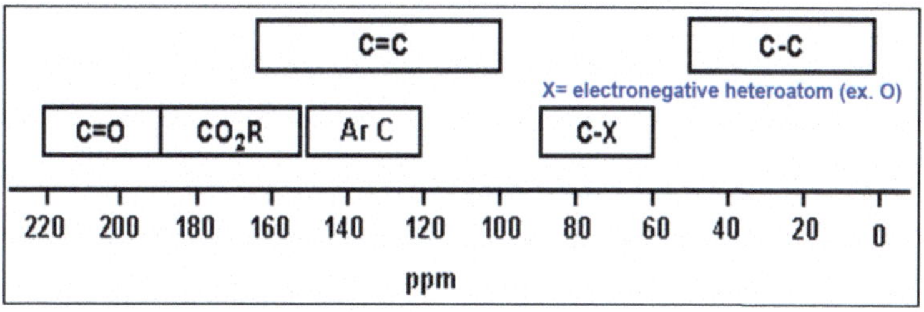

Fig. 2.55 Chemical shifts of compounds in ^{13}C-NMR

(x) peak overlap is much less common than for ^{1}H-NMR, which makes it easier to determine how many types of C are present.

2.9.1.3 Correlation Spectroscopy

Correlation spectroscopy is one of several types of 2D-NMR. This type of NMR analysis is best known by its acronym **COS**. Other types of 2D-NMR include J spectroscopy, exchange spectroscopy (**EXSY**), nuclear Overhauser effect spectroscopy (**NOESY**), total correlation spectroscopy (**TOCSY**) and heteronuclear correlation experiments, such as **HSQC**, **HMQC** and **HMBC**. 2D-NMR spectra provide more information about a molecule than 1D-NMR spectra and are especially useful for determining the structure of a highly complex molecule.

2.9.2 NMR Applications

Magnetic Resonance spectroscopy is notably the most important for the investigation of molecular structures, if there are very close relationships between the data obtained by ^{1}H-NMR and the arrangement of protons in the molecule of interest. Table 2.7 summarizes the type of information obtained with NMR.

This information often allows proposing a structure for an unknown molecule, based solely on its spectrum, reducing the number of possibilities.

Nuclear magnetic resonance spectroscopy is a technique that has been widely used due to the possibility of studying a large number of nuclei in all phases (gas, liquid and solid), even at high temperatures, in a non-destructive and safe way. Some of its applications include: verification of the purity of mineral raw materials; quality control of chemical products; investigation of chemical reactions and investigation of unknown substances. Figure 2.56 presents a comparison between NMR applied to samples of different natures.

In the specific case of analysis of mineral raw materials, NMR is used when the analysis parameters are characteristic of the behavior of the nuclear *spin* of the atoms involved in the chemical bonds present in the solid sample. In conventional NMR applied to liquid solution samples, these interactions would lead to a significant broadening of the spectral lines. A variety of techniques allows for high-resolution conditions, which can, at least for ^{13}C spectra, be comparable to spectra obtained in conventional NMR of liquid samples.

Two important concepts for high resolution in solid sample NMR spectroscopy are molecular orientation limitation by sample orientation, and the reduction of anisotropic nuclear magnetic interactions in sample rotation. This latter, rapid rotations of the sample around the **magic angle**, is a very important method, when the system is composed of *spin* 1/2 nuclei. The magic angle is a technique that allows to eliminate or reduce dipolar couplings, thus increasing the spectral resolution. Figure 2.57 presents typical spectra observed in solid NMR.

Table 2.7 Types of information about molecular structure, obtained by NMR spectrometry

Information contained in the spectrum	Arrangement of protons
Number of resonance lines	Number of equivalent proton groups
Position of resonance signals	Type of proton groups
Relative intensity of resonance signals	Number of protons causing the signal
Fine structure of resonance signals	Position of proton groups relative to each other

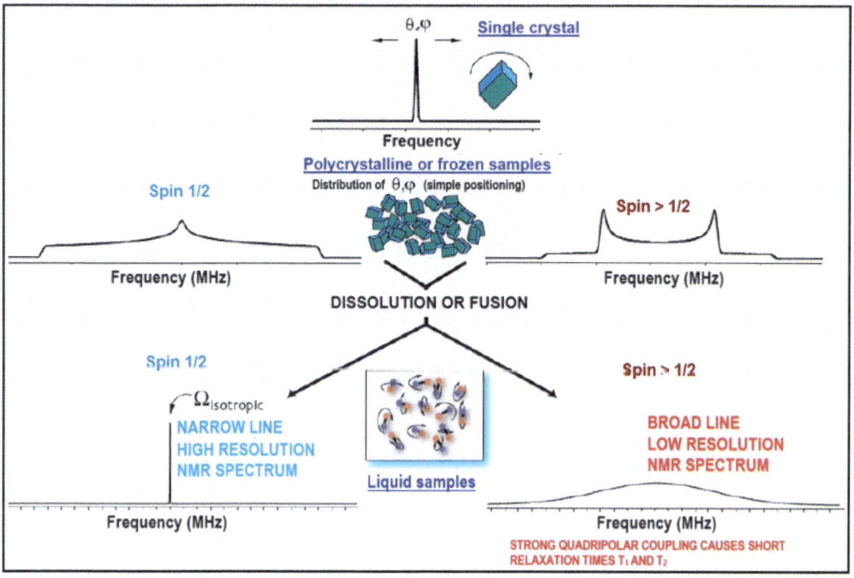

Fig. 2.56 Schematic comparison between NMR applied to samples of different natures

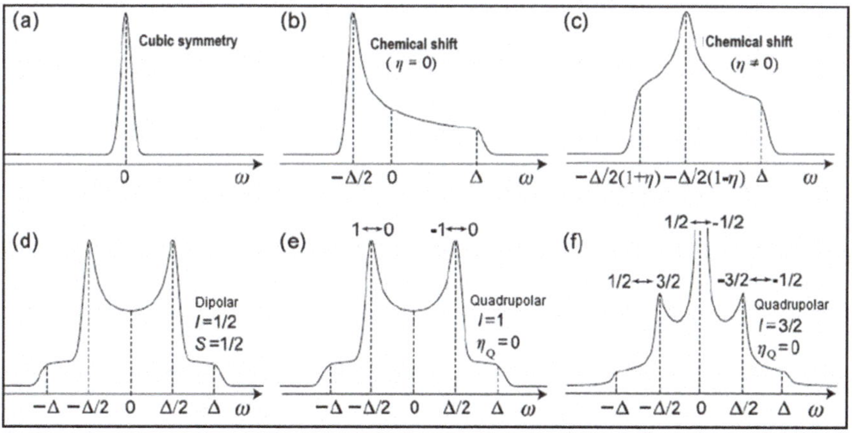

Fig. 2.57 Typical spectra observed in solid NMR: **a** cubic symmetry; **b** axial and **c** non-axial symmetry; **d** interaction between 2 spins ½ (I and S); quadrupolar interactions for spins, **e** 1 and **f** 3/2

2.9.3 *Sample Preparation in NMR*

Samples for NMR are usually prepared in 5 mm diameter glass tubes. Whenever possible, deuterated solvents are used, because deuterium blocks frequency fields,

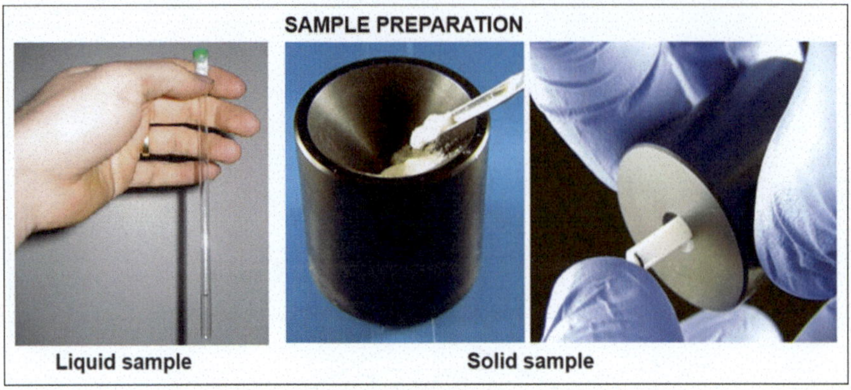

Fig. 2.58 Sample holders used for liquid samples and for solid samples

making spectral resolution easier and faster. The sample needs to be sufficiently soluble, a height of 4.5–5.0 cm is recommended.

For solid samples, if there is not enough sample, it can be diluted in an inorganic solvent, such as aluminum sulfate or calcium sulfate. Solid samples should be ground. The powder after grinding should be compressed with a spatula into a sample holder. Figure 2.58 illustrates the types of sample holders used for liquid samples and for solid samples.

2.9.4 Sources of Error in NMR

The sources of error in NMR are varied and can arise from the insensitivity of the method, as well as due to the presence of multiple components in the sample. The main source is undoubtedly error in the interpretation of the results, due to incorrect reading of the analysis parameters.

Also, the control of the concentration of the solutions must be strict. For high molecular weight samples, the concentration should be increased, avoiding, however, excessive concentration which can increase the viscosity contributing to a low resolution (Fig. 2.59).

2.10 X-Ray Diffraction

The analysis of the mineralogical composition of mineral raw materials is usually carried out by X-ray diffraction (XRD). This technique provides predominantly qualitative, but also quantitative, information about the crystalline structure of mono or

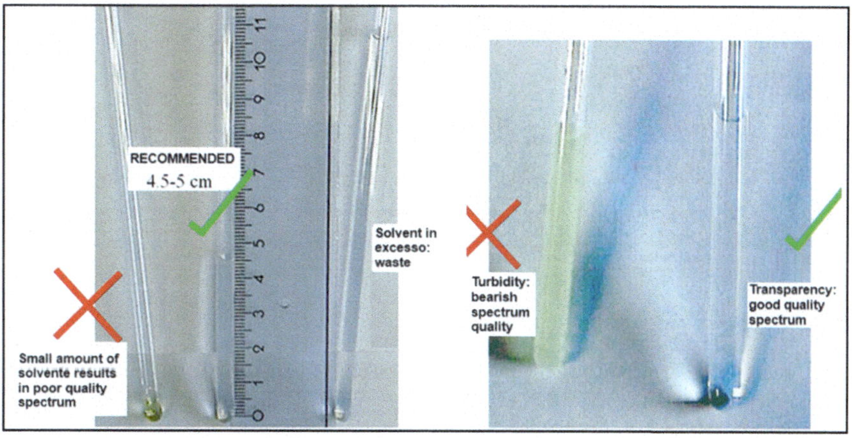

Fig. 2.59 Preparation of liquid sample in NMR: precautions to avoid errors in spectrum reading

polyphasic samples in a simple, relatively fast and reliable way, as the spectrum obtained is unique for a given crystalline phase.

The acquisition of information by XRD about the mineralogical composition of a mineral raw material is possible because, in most solids (crystals), the atoms arrange themselves in crystalline planes separated from each other by distances of the same order of magnitude as the wavelengths of X-rays (between 0.70 and 2.30 Å) with the systematic repetition planes of the crystalline lattice. When an X-ray beam hits a crystal, it interacts with the present atoms, originating the phenomenon of diffraction. Figure 2.60 presents the division of the regions of the electromagnetic spectrum, by wavelength and frequency. In detail, the range of X-ray wavelengths for XRD.

The determination of the phases is made by comparing the obtained spectrum with cataloged data, maintained by the JCPDS (Joint Committee on Powder Diffraction Standards), considering the interplanar distance in the crystalline lattice for a known angle of incidence (θ). Figure 2.61 presents characteristics of X-ray diffraction.

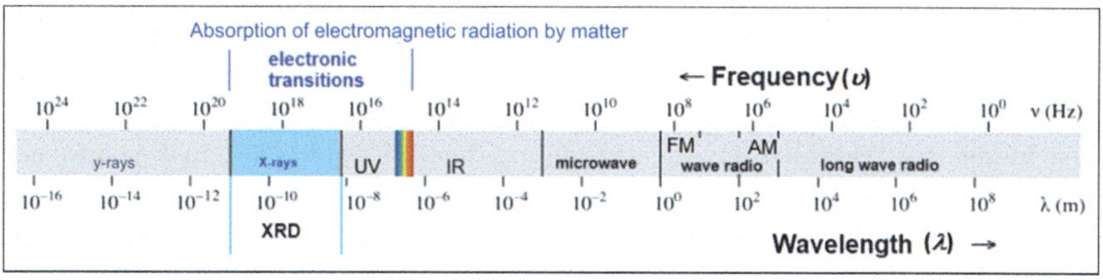

Fig. 2.60 Division of the regions of the electromagnetic spectrum, by wavelength and frequency. In detail, the range of X-ray wavelengths for XRD

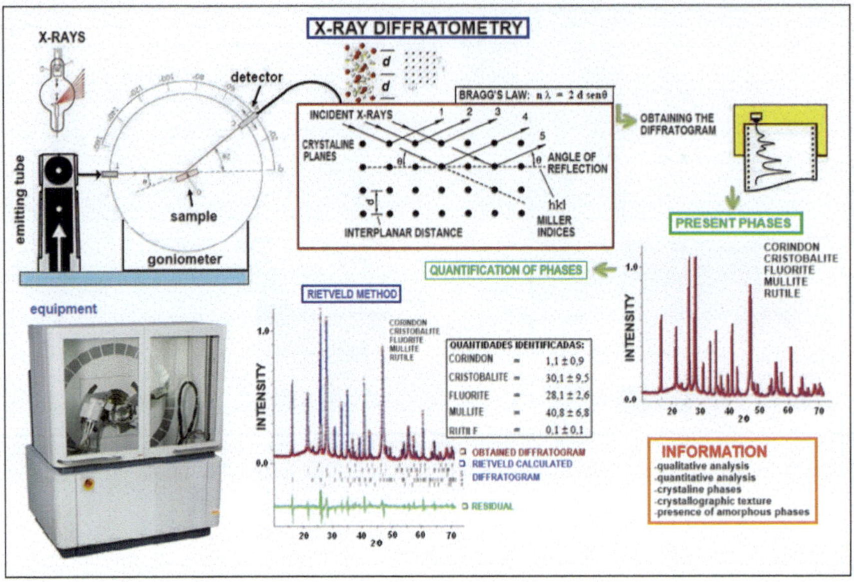

Fig. 2.61 Characteristics of X-ray diffraction applied to mineral raw materials

2.10.1 Principle of X-Ray Diffraction

X-rays are generated when a high kinetic energy particle is rapidly decelerated. Figure 2.62 illustrates the phenomenon at the atomic level. The way the X-ray spectrum behaves is explained through transitions of atomic energy levels. For each different energy level transition, a different wavelength is emitted. For example, K α_1 radiation is produced when an electron transitions from the L_{III} layer to the K layer. The energy of the emitted photon equals the energy difference between the two layers. As the energy for each level varies with the atomic element (target), each type of target produces characteristic radiations at different wavelengths, as can be observed in Table 2.8, for the materials used in X-ray tubes.

The scattering of X-rays can be **coherent** or **incoherent**. In the **coherent** scattering, the scattered wave has a defined direction, same phase, and same energy in relation to the incident wave, it is an elastic collision. In the **incoherent** scattering, the scattered wave does not have a defined direction, does not maintain the phase nor the energy (**Compton Effect**). The collision is inelastic, and the energy transforms into temperature (atom vibration).

From the point of view of wave physics, it can be said that the electromagnetic wave is instantly absorbed by the electron and reemitted, each electron therefore acts as a center of X-ray emission. For constructive interference of the scattered waves to occur, it is necessary to satisfy the following condition given by Eq. 2.10, known as **Bragg's Law**.

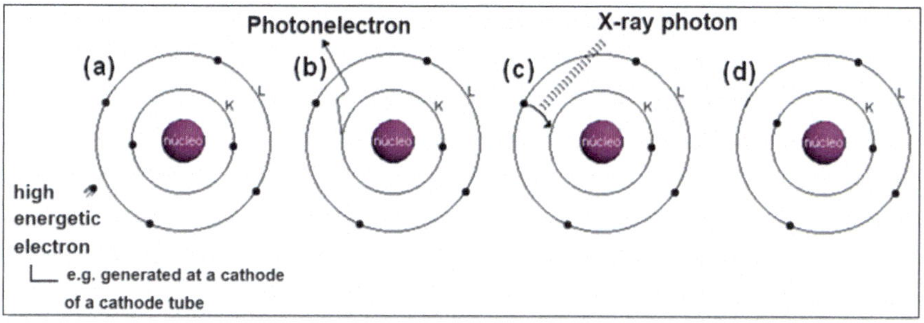

Fig. 2.62 Production of X-rays at the atomic level. **a** an electron accelerated hits the target (for example, this high-energy electron can be generated at the cathode of a cathode tube); **b** an electron from the K layer of an atom of the material is released in the form of a photoelectron, creating a vacancy in this layer; **c** to fill this vacancy an electron from the outermost layer moves to the *K* layer, releasing energy in the form of an X-ray photon and **d** the atom assumes the new configuration. The energy of the photon corresponds to the energy difference between the two layers

Table 2.8 Characteristic radiations of the main materials used in X-ray tubes

Element	K α_1 (Å)	K β_1 (Å)
Cu	1,54,056	1,39,221
Mo	0,70,930	0,63,228
Cr	2,28,970	2,08,487
Co	1,78,896	1,62,079
W	0,20,901	0,18,437
Ni	1,65,791	1,50,013
Fe	1,93,604	1,75,661

$$n\lambda = 2d \sin\theta \tag{2.10}$$

where:

n = integer number;
λ = wavelength of the used X-ray source;
d = interplanar distance;
θ = angle of reflection.

If the atoms that generate this scattering are arranged in a crystalline structure (Fig. 2.63), with distances close to the wavelength of the incident radiation, it can be verified that the phase relations between the scatterings become periodic and the diffraction effects can be observed at various angles.

The configuration of an X-ray diffractometer depends on the method to be used in the identification of samples. Figure 2.64 schematically presents 3 different techniques of analysis by X-ray diffractometry.

(i) **Crystal Rotation Method**: a single crystal is arranged with one of its crystallographic axes perpendicular to the X-ray beam. A cylindrical film is placed

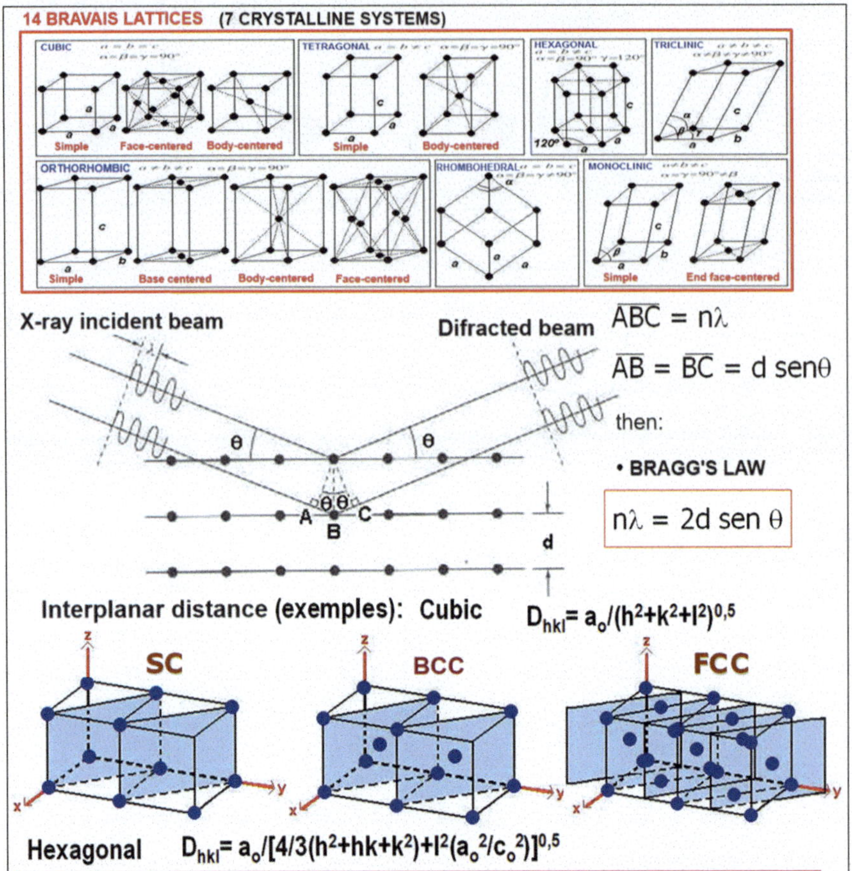

Crystalline structures	Diffraction does not occur	Diffraction occurs
BCC FCC HC	h+k+l=even number h, k, l (odd and even) h+2k= l even (n is na integer)	h+k+l= odd number r h, k, l (or even or odd) all other cases

Fig. 2.63 Crystalline structures and X-ray diffraction at the planar level, according to **Bragg's Law**

around it and the crystal is rotated around the chosen axis. At some point, Bragg's Law will be satisfied, and a diffracted beam will be formed.

(ii) **The Laue method** is mainly used to determine the orientation of large single crystals. White radiation is reflected or transmitted through a fixed crystal. The diffracted beams form curves on the film. The Bragg angle is fixed for each set of planes in the crystal, which refracts the determined wavelength of the white radiation that satisfies Bragg's law for the values of d and θ involved. Experimentally, there are two variants of the Laue method: transmission or backscattering, as shown in Fig. 2.64.

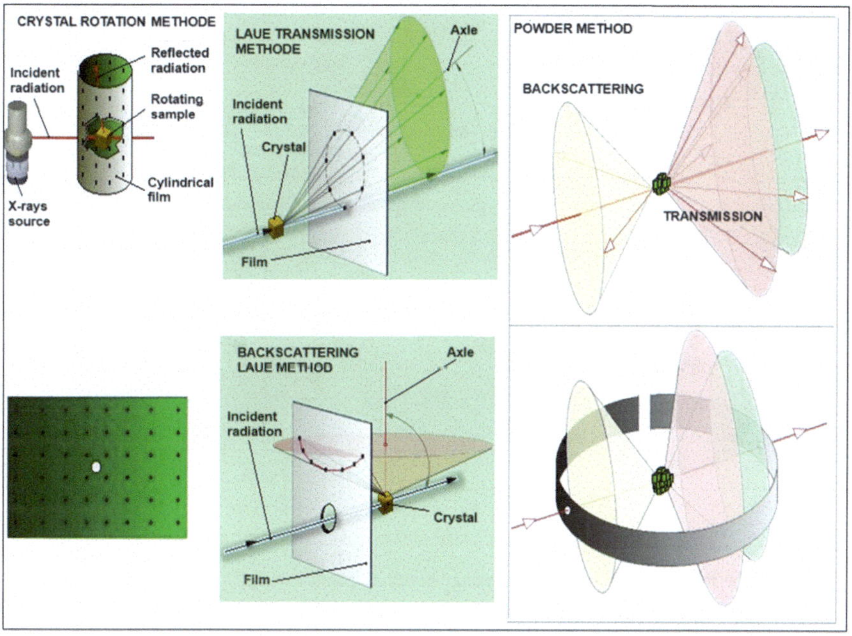

Fig. 2.64 Three different methods of analysis by X-ray diffractometry: **crystal rotation method**; **Laue method** and the **powder method**

(iii) The **powder method** is used to determine the value of the lattice parameters accurately. A powder sample forms diffracted beams in the configuration of continuous cones, as it is made up of numerous crystals each with its own orientation. The lines are seen as arcs on the film. The sample is placed in the center of a cylindrical device, called the Debye-Sherrer Chamber (Fig. 2.65).

In the case of samples of mineral raw materials, the powder method is used where the incident wavelength is fixed, and the angle of reflection is variable. The basic configuration of an X-ray diffractometer, for the powder method, is schematized in Fig. 2.64. It is observed that the sample is located in the center, from which the X-ray source and the slit of equidistant detector (Bragg's circle). In the most common configuration, called $\theta/2\theta$, the sample moves at an angular speed θ, and the detector at double this speed, so that it is always detecting the counts reflected in the sample plane. The reading that is made are counts reflected at a certain angle θ, converted to interplanar distances d by the Bragg equation.

The determination of the phases is made by comparison between the radiation diffracted by a sample as a function of its crystalline structure. The radiation from an X-ray source passes through the sample, being then detected and compared with cataloged data. By similarity, considering the interplanar distance for a known incidence angle θ, it is then defined which are the crystallographic (mineralogical)

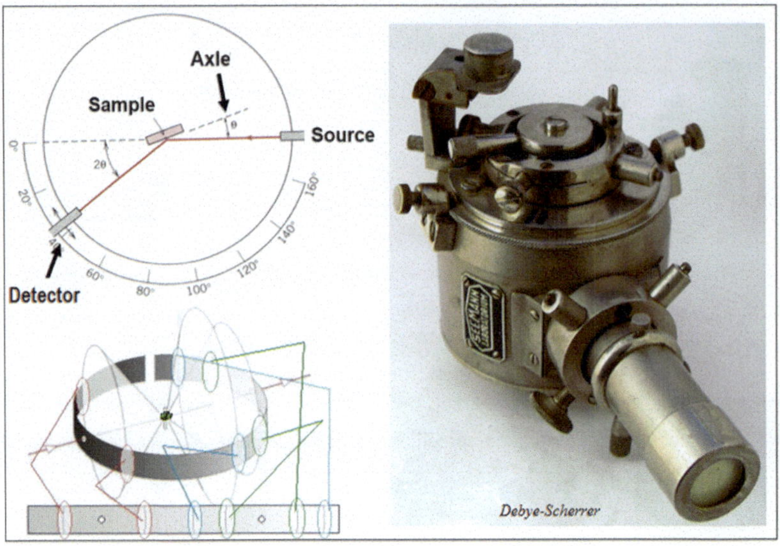

Fig. 2.65 Schematic representation of a basic X-ray diffractometer for the powder method and Debye–Scherrer chamber

phases present in the sample. Figure 2.66 presents the characteristic diffractograms of amorphous SiO_2 and in different crystalline structures.

2.10.2 Applications of X-Ray Diffractometry

X-ray diffraction is extensively used for the definition, for example, of clay minerals in refractory clay. The analysis can be done on the entire raw material, indicating, therefore, the presence of other constituents besides the clay minerals (such as quartz) or only the fraction less than 2 μm, consisting only of clay minerals.

The estimation of the size of the crystallites through X-ray diffraction by mathematical treatment of the peaks (peak width) by the method called ***single-line*** has gained increasing importance due to the rapid expansion in the area of research of nanostructured materials. This technique is being widely used in determining the size of the crystallite in substitution to the analysis by transmission electron microscopy, as the latter requires a delicate and meticulous preparation of the samples and the interpretation of the results requires a high level of knowledge and practice, in addition to incurring problems of sample representativeness (the analyzed region is typically a few nanometers) and the time required for the analysis is considerably longer than when performed by X-ray diffraction.

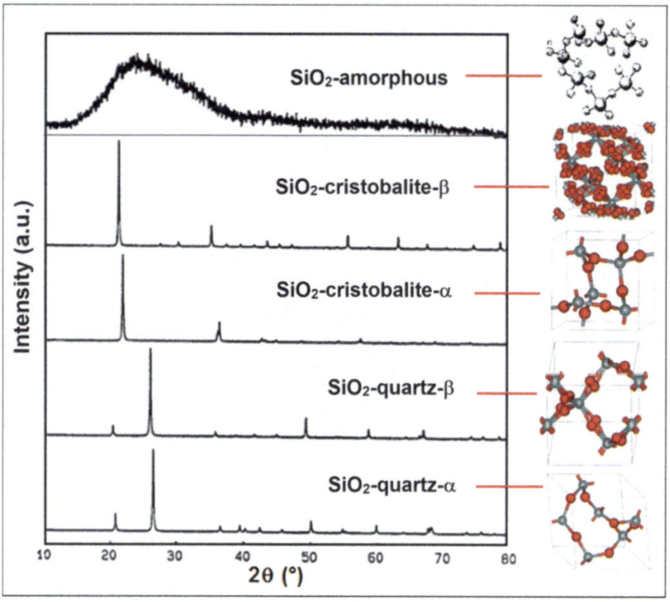

Fig. 2.66 Characteristic diffractograms of amorphous SiO_2 and in different crystalline structures

2.10.2.1 Quantitative Analysis of X-Ray Diffraction (Rietveld Method)

The **Rietveld method** is a procedure for quantitative phase analysis, with recognized precision. This method is based on the comparison of the observed X-ray diffraction pattern, with a calculated pattern from the structure data (crystallographic parameters) of the existing phases. The quantification by the Rietveld method is based on three initial considerations:

(i) each crystalline structure has its own diffraction spectrum characterized by the positions and intensities of each diffraction peak;
(ii) the superposition of diffraction spectra is done by simple addition, without interference; and;
(iii) the integral of the surface of the spectrum of each phase is proportional to the percentage of the phase in the mixture.

The quantitative data are deduced from scale factors, after the superposition and minimization of differences between the measured and calculated spectra, the latter obtained from modeling of each phase in the mixture. Three groups of parameters should be included in the modeling:

(i) instrumental parameters (equipment zero correction);
(ii) structural parameters, such as space group, unit cell parameters, atom positions in the lattice and their occupancy, absorption and thermal factors;
(iii) crystallinity parameters, such as peak asymmetry, preferred orientation, half-width and peak shape.

For accuracy of the technique, care should be taken with the preferred orientation of phases, obtaining a high count and the presence of phases with different microabsorption from the others. This last case usually happens with the goethite, hematite and magnetite phases (for the Cu $K\alpha$ radiation used). This effect can be eliminated by using more appropriate radiation, or minimized by more intense pulverization of the sample. It is estimated that the differential microabsorption interferes less, from 5 μm particle size. In this case, care must be taken with overgrinding, which can affect the crystallinity of the minerals.

The Rietveld method also allows quantifying the amorphous material, this is important in the case of lateritic ores, derived from the intense action of weathering as is the case with most Brazilian ores. For this, it is necessary to add a known amount of an internal standard, generally 10–15% by mass. The internal standard must be perfectly crystalline, not be among the phases identified in the original sample, and preferably be easy to refine (the most used are silicon, and corundum).

An illustrative example is presented in Fig. 2.67. The X-ray diffraction spectrum of a Brazilian bauxite is presented, where gibbsite, kaolinite, goethite and muscovite were identified. Below the collected spectrum, the calculated spectra are presented, after refinement, and the difference between both. For the identification of the amorphous phase, it was necessary to add 15% (by mass) of fluorite, disregarded in the final quantification of the sample.

Case Stud. The authors König, Pöllmann and Angélica (2002) studied the mineralogy of iron ores (hematite) from two different sources (Itabira, Brazil and Cumberland, United Kingdom) and application of the Rietveld method in a sample of iron ore manufactured in the laboratory, containing 4 mineral phases: hematite, magnetite, goethite and quartz.

Figures 2.68 and 2.69 present samples of the 2 different occurrences of hematite investigated.

Minerals in the form of lamellar aggregates, such as hematite, present strong texture effects in X-ray diffraction. These effects are more pronounced for the (006) planes (Figs. 2.68 and 2.69, red arrow). The diffraction intensity on these planes will depend on some factors, such as, for example:

(i) The reduction of texture effects by variation in grinding time and type of mill used;
(ii) The use of different counting times, X-ray tubes, voltage and current intensities in the X-ray diffractometer.

Figure 2.70 illustrates the dependence of the factors **full width at half maximum** (FWHM), intensity and peak area with the duration of grinding. The parameters were calculated with the WINFIT program. In the analysis of Fig. 2.69, it can be observed, from trend lines, that with the increase in the duration of grinding, the intensities and the areas of the peaks decrease exponentially, while the FWHM factor increases.

The quantification of phases through the **Rietveld Method** for the mixture of three phases (hematite, magnetite, goethite) is presented in Fig. 2.71. The mixture with 75% hematite (Itabira), 10% magnetite and 15% of goethite was prepared and analyzed four times. The relative standard deviation (RSD)—which depends on the

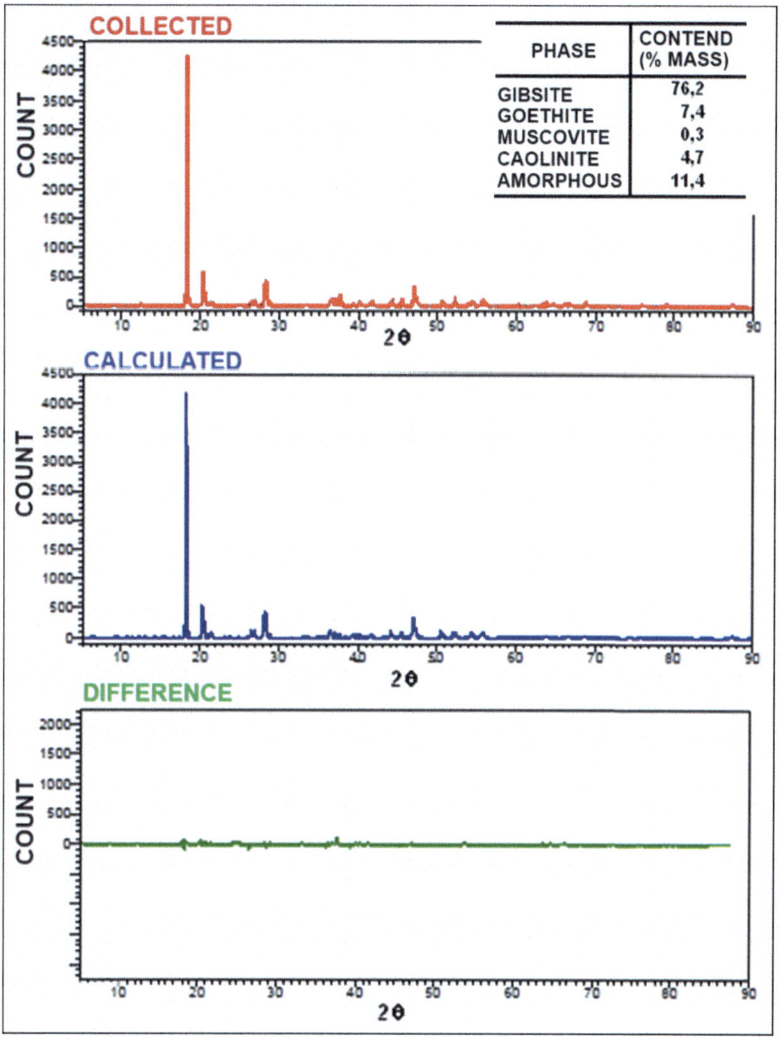

Fig. 2.67 Quantitative analysis by the Rietveld method for a Brazilian bauxite: X-ray spectra measured and calculated by the Rietveld method, and the difference between the two

variety of hematite—is 0.8% for the hematite, 3.0% for the magnetite and 2.6% for the goethite.

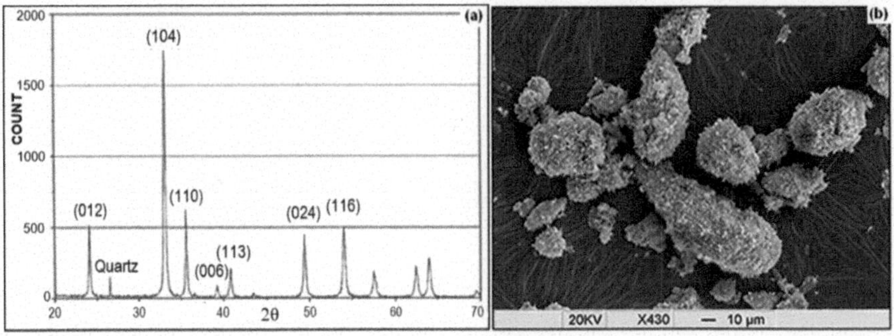

Fig. 2.68 X-ray diffractogram and SEM of radial aggregates of hematite from Cumberland (United Kingdom)

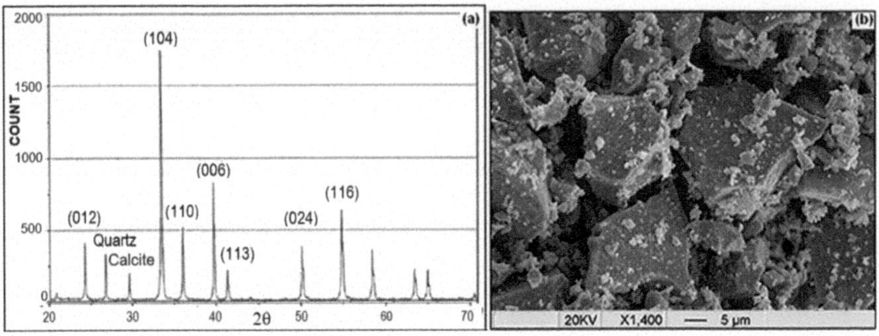

Fig. 2.69 X-ray diffractogram and SEM of lamellar hematite from Itabira (Brazil)

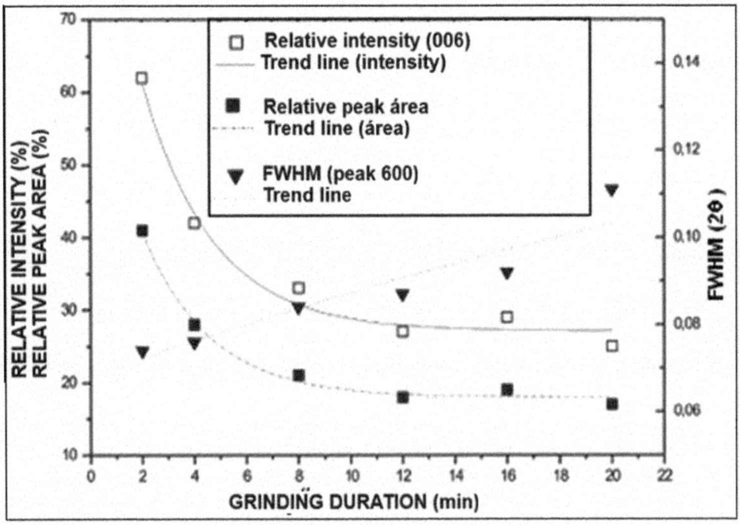

Fig. 2.70 Dependence of the factors FWHM, relative intensity and relative peak area with the duration of grinding

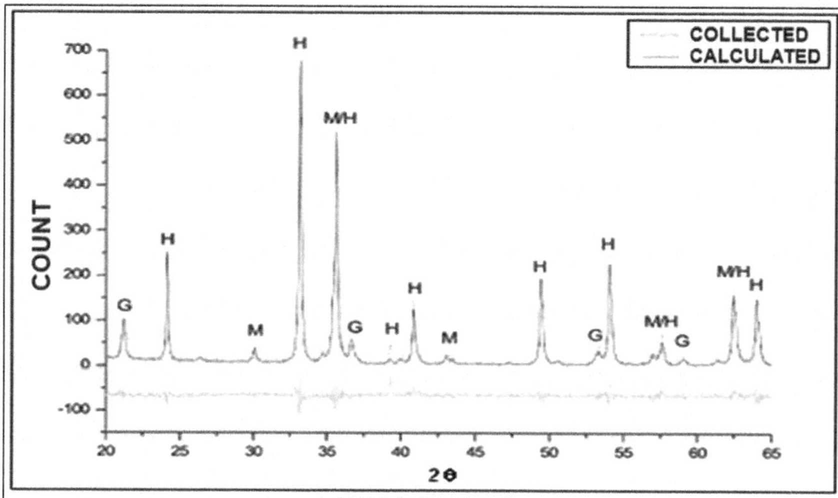

QUANTITICATION

Mistura	Hematite (75%)	Magnetite (10%)	Goethite (15%)
Quantification 1	75.0	10.1	14.9
Quantification 2	74.3	10.1	15.6
Quantification 3	75.6	9.8	14.6
Quantification 4	74.4	10.5	15.1
Average	74.8	10.1	15.1
Standard-desviation	0.6	0.3	0.4

Fig. 2.71 Patterns of a mixture 75(74.3)% hematite (H) + 15(15.6)% goethite (G) + 10(10.1)% magnetite (M), (2θ Cu *Ka*). Below the graph are shown the results of the quantifications for four mixtures

2.10.3 Sample Preparation for XRD

The preparation of samples for XRD analysis of raw mineral materials is quite simple for the powder method. The sample should be comminuted to a particle size, as a rule, below 60# (200 μm). This preparation should be standardized so that significant errors are not introduced in the identification of mineralogical phases.

2.10.4 Source of Error in XRD

The interpretation of spectra of samples with good crystallinity and monophasic, or simple mixtures, is very easy, and usually the search tools of the *software* for interpretation of diffraction spectra more modern correctly identify the phases present. As a greater diversity of minerals coexist in the sample, the complexity of the structures increases. Similarly, when it is necessary to identify minerals present in small amounts, XRD becomes more susceptible to errors in the interpretation of diffractograms.

The position of the peaks (angle θ, or interplanar distance after calculation by Bragg's Law) of a certain mineral should not change, but in practice there may be small deviations, mainly due to variation in its composition or any poor sample preparation.

The relative intensities of the peaks, on the other hand, are greatly affected by preferential orientation in the preparation of the samples, by the superposition of peaks of different phases, by the variation of the composition of the minerals, and by other factors. In addition, the intensity (and the width of the peaks) depends a lot on the crystallinity of the phases. The simple comparison of height or the integral of the area of the peaks does not represent quantification, and does not allow quantitative comparison of the phases.

The spectrum in Fig. 2.72 presents two different diffractograms for a simple comparison. In the upper diffraction spectrum, it is graphically represented and the pattern matches the mineral, in this case quartz. In the lower diffraction spectrum, a mixture of various minerals is noted, with strong overlap of some of the main peaks, and whose interpretation, with certainty, requires additional data (optical or scanning electron microscopy, for example).

A very deleterious fact to the analysis of weathered ore samples, typical of more tropical climates, is the possible presence of phases with high Fe content, including amorphous or poorly crystallized (such as limonite). Therefore, if the equipment uses a copper tube for the generation of X-rays, it is necessary to use secondary monochromators to avoid a very low signal/noise ratio, which masks most of the peaks. This is because Fe presents strong fluorescence when excited by Cu radiation. As weathered minerals are often also poorly crystallized, identification is even more laborious.

Another problem can occur when the structure is affected by its bombardment by radioactive elements (metamictization), very common, for example, in zircon (or zirconite) or in pyrochlore. In this case, the structure can eventually recompose itself, to the point of allowing correct identification, if the mineral is heated to high temperatures (on the order of 2/3 of its melting point, around 850 °C in the examples), for 24 h.

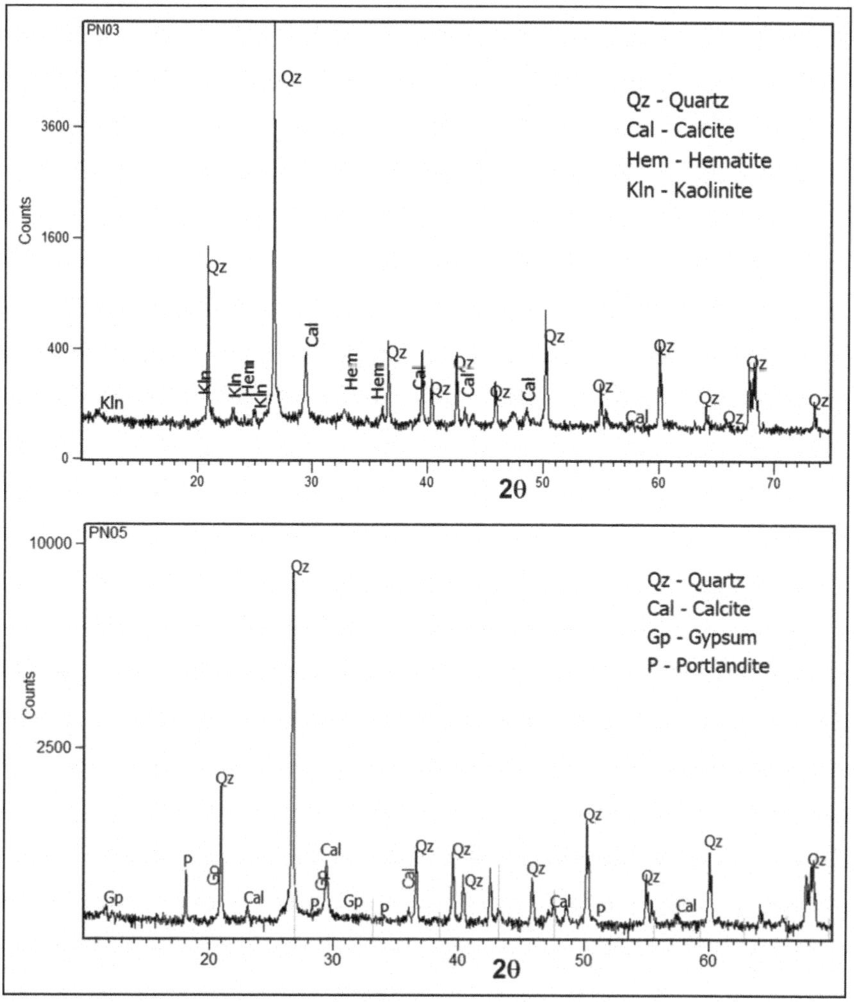

Fig. 2.72 X-ray diffraction spectrum (solid line). Upper spectrum: quartz sample, with position and relative count to the tabulated standard of the quartz, calcite, hematite, and kaolinite. In the lower spectrum, sample of complex mineralogy, with peak overlap. Identified minerals: quartz, calcite, gypsum, and portlandite. (https://gmga.com.br/03-xrd-mineralogical-characterization-of-mortars-from-the-sao-pedro-nolasco-fort-in-belem-para/)

Chapter 3
Particle Size Analysis

3.1 Introduction

The determination of the particle size of mineral raw materials is of fundamental importance in industrial processes. Particle size analysis is essential for physical processes aimed at extraction and synthesis. Prior knowledge of the particle size distribution of the particles to be treated is of great importance, as this information will define the operating conditions and the equipment that will lead to the desired particle size of the product, as well as being used for the calculation of the energy expended to produce it. It will also serve as the basis for controlling the production of fine particles in fragmentation processes, for defining the degree of liberation and the distribution of chemical and mineralogical elements in each specific size range and for quality control of the product according to market specifications.

The specification of particle size requires the definition of what is understood by the size of the material's particles. Five types of particulate solids are distinguished by size, although this distinction is not very clear, according to Table 3.1.

The values referred to in a granulometric analysis usually use sieves of different openings as a reference, which are standardized internationally. Table 3.2 presents a relationship between sieve openings (in mm) and standardization, according to the BSS, Tyler, and US scales. Each sieve has several **openings per linear inch** called ***mesh***. Therefore, the larger the *mesh*, the greater the number of openings and, consequently, the smaller the particle must be to pass through the opening. Thus, for coarse materials, low *mesh* sieves are used and for fine materials, sieves with a larger *mesh* are used.

The **granulometric analysis** is represented in the form of typical graphs where the granulometric fractions and the frequency of distribution are related. The granulometric distribution can also be represented by the parameters D_{10}, D_{50}, D_{90} and $D_{average}$. Figure 3.1 presents the efficiency intervals in the granulometric analysis of raw materials and parameters for representing the granulometric distribution of raw materials.

C. P. Bergmann, *Techniques for Analysis of Mineral Raw Materials*, Topics in Mining, Metallurgy and Materials Engineering, https://doi.org/10.1007/978-3-031-58985-0_3

Table 3.1 Types of particulate solids according to their particle size

Type of solid	Particle size
Powders	Particles from 1 μm to 0.5 mm
Granular solids	From 0.5 to 10 mm
Small blocks	Particles from 1 to 5 cm
Medium blocks	Particles from 5 to 15 cm
Large blocks	Particles larger than 15 cm

Table 3.2 Relationship between sieve openings (in mm) and standardization according to the BSS, Tyler, and US scales

Opening size (mm)	BSS	Tyler	US
4.75	–	4	4
3.35	5	6	6
2.81	6	7	7
2.38	7	8	8
2.00	8	9	10
1.68	10	10	12
1.40	12	12	14
1.20	14	14	16
1.00	16	16	18
0.853	18	20	20
0.710	22	24	25
0.599	25	28	30
0.500	30	32	35
0.422	36	35	40
0.354	44	42	45
0.297	52	48	50
0.251	60	60	60
0.211	72	65	70
0.178	85	80	80
0.152	100	100	100
0.125	120	115	120
0.104	150	150	140
0.089	170	170	170
0.075	200	200	200
0.066	240	250	230
0.053	300	270	270
0.044	350	325	325
0.037	440	400	400

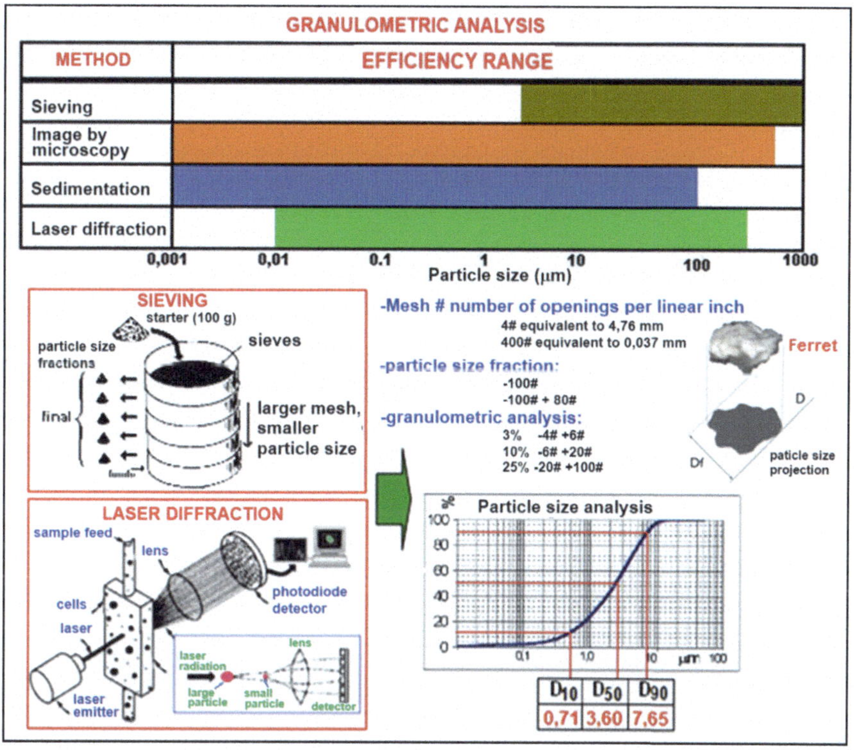

Fig. 3.1 Granulometric analysis methods of raw materials, efficiency limitations and type of information

The issue of determining a representative size of a mineral raw material is not a simple solution. Representing the size of a particle with a single number is quite limiting in relation to its granulometry. Measuring, for example, the size of a rod-shaped particle with a linear scale is relatively complex. Any measure based on a single dimension of the particle would be limited as information.

In the quality control of an industrial process, involving finely divided particles, a single number representing the granulometry would be the ideal situation. Thus, the question is posed: how to describe a three-dimensional solid using only one number.

The Equivalent Sphere theory advocates the possibility of describing a property with a single number, like the equivalent of a sphere of diameter *d*. A sphere of 50 μm leaves no doubt about the size it refers to. The same does not occur with other geometric shapes, such as the cube: 50 μm can be the edge or diagonal. Already properties like weight, volume and surface area can be described with a single number. Thus, the weight of a three-dimensional particle can be converted into the weight of a sphere (weight = $4/3\ \pi r^3 .\ \rho$), allowing the calculation of a single number (2r) for the diameter of a sphere that has the same weight as the particle.

Ferret proposed a minimum diameter (Ferret diameter—D_f), which would be the diameter of a circle with the same area as the two-dimensional projection of a particle, as representative of the particle's granulometry.

As can be seen, the results of a measurement lose their value if comparing particle sizes using different techniques (or references) in the measurement. This becomes quite critical, for example, in an industrial process that does not adopt a single methodology for granulometric analysis. However, it is not always possible to have a single particle size standard, as there are limitations of the analysis techniques in function of the granulometry of the particles in the sample of interest.

Among the techniques for the granulometric analysis of an industrial raw material, the most conventional is sieving, separating the raw material into granulometric ranges. However, the efficiency of this separation is limited to larger, relatively coarse sizes.

For smaller granulometry ranges, micro sieves can be used, but the efficiency, cost, time, and precision drop significantly. For these smaller ranges, cyclone separation and image analysis by electron microscopy can be used. However, the first is falling out of use for being quite laborious (separation, weighing, cleaning of cyclones) and the second for lacking a greater precision in sampling and representativeness.

Lately, the technique of laser radiation diffraction, used by laser granulometers, has stood out as a very fast and efficient technique for granulometries in the range of a few mm to tens of nanometers. In a few minutes, the granulometric distribution of the sample (a few grams) of interest can be obtained. The analysis can be performed on dry samples or in a liquid medium.

3.2 Granulometric Analysis by Sieving

Granulometric analysis by sieving is the most conventional method, characterized by direct measurements and used in a range of sizes up to about 40 μm. The material retained on the sieve screen is called *oversize* and the passing, *undersize*. In wet sieving, water is added to the material to be sieved with the purpose of facilitating the passage of fines through the sieving screen. Figure 3.2 summarizes the main aspects related to the sieving of mineral raw materials.

The sieving surface of the sieves, the mesh, is made up of square openings, formed by perpendicularly braided wires. The sieves used constitute a standardized series, according to Table 3.1, whose openings are related to each other by a geometric progression, allowing the comparison of the classification results.

The mass of material to be tested in a series of sieves should not allow that at the end of the sieving the mass retained on each sieve exceeds the equivalent of 3 layers of particles. A very large number of particle layers can prevent those located in higher layers from reaching the sieve mesh and, in this way, would not be tested. Thus, for granulometric analysis to be possible it is necessary that the sieves retain a mass smaller than that estimated by the **Gaudin formula**, given by Eq. 3.1.

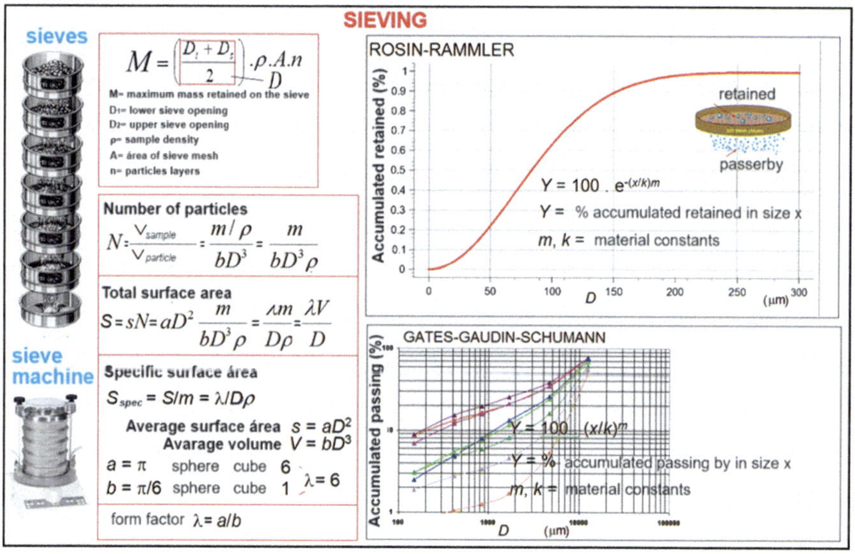

Fig. 3.2 Main aspects related to the sieving of mineral raw materials

$$M = \rho.A.n.\frac{(Di + Ds)}{2} \tag{3.1}$$

where:

M = maximum mass retained on each sieve (g);
D_i = opening of the sieve in question (cm);
D_s = opening of the sieve immediately above the scale (cm);
ρ = density of the sample to be tested (g/cm^3);
A = area of the sieve mesh (cm^2);
n = layers of particles.

The granulometric distribution of a sample can be described mathematically by various empirical equations, found in literature and, among these, the Gates-Gaudin-Schumann equation (Eq. 3.2) and the Rosin–Rammler equation (Eq. 3.3) stand out. Generally, these equations are used in an attempt to obtain a linear distribution between the data obtained during classification.

$$Y = 100\,.\,(x/k)^m \tag{3.2}$$

where:

Y = % cumulative passing at size x;
m, k = material constants.

$$Y = 100.\mathrm{e}^{-(x/k)m} \tag{3.3}$$

Y = % cumulative retained at size x.

m, k = material constants.

The particle size can also be provided in a simpler way as understood between the average of the mesh that retained (D_i) and the immediately previous (D_s). The arithmetic mean of these mesh openings will serve to characterize the physical size of the particle (D), according to Eq. 3.4.

$$D = \frac{(Di + Ds)}{2} \tag{3.4}$$

In this way, important characteristics of the material can be estimated based on D, such as average surface area (Eq. 3.5), average volume (Eq. 3.6) and shape factor λ (Eq. 3.7), number of particles in the sample N (Eq. 3.8), total surface area S (Eq. 3.9) and the specific surface area (Eq. 3.10).

$$s = aD^2 \tag{3.5}$$

where:

$a = \pi$ for the sphere and 6 for the cube.

$$V = bD^3 \tag{3.6}$$

where:

$b = \pi/6$ for the sphere and 1 for the cube.

$$\lambda = a/b \tag{3.7}$$

For cubes and spheres, $\lambda = 6$. Many grinding operation products have $\lambda = 10.5$. For lamellar particles of mica, $\lambda \approx 55$.

$$N = \frac{Vamostra}{Vamostra} = \frac{m}{\rho bD3} \tag{3.8}$$

where:

ρ = material density;

$$S = sN = aD2\frac{m}{\rho bD3} = \frac{\lambda m}{D\rho} = \frac{\lambda V}{D} \tag{3.9}$$

$$Sesp. = \frac{S}{m} = \frac{\lambda}{D\rho} \tag{3.10}$$

3.3 Granulometric Analysis by Laser Diffraction

The particle size distribution analysis by LASER (Light Amplification by Stimulated Emission of Radiation) is a high-resolution method that allows the evaluation of particles much smaller than the traditional method, sieving.

A beam of light, when it hits a particle, can be reflected, absorbed, diffracted or refracted. Normally, these phenomena occur simultaneously, generating a scattering pattern with characteristic maximum and minimum intensity values.

The operating principle of a granulometry equipment by LASER diffraction is the analysis of the scattering pattern from the sum of the pattern generated by each particle present in the sample (Fig. 3.3). Through mathematical treatment on the resultant, it is possible to obtain information about the scattering pattern of each individual particle, and through inversion, data about their sizes are obtained. For an accurate analysis, the scatterings must be restricted to a particle, without interacting with another, before reaching the detector. To avoid this phenomenon, the sample concentration must be controlled.

By the assumptions of the Fraunhofer approximation, used in the estimate of granulometry, the particles are considered opaque spheres, with diameter much larger

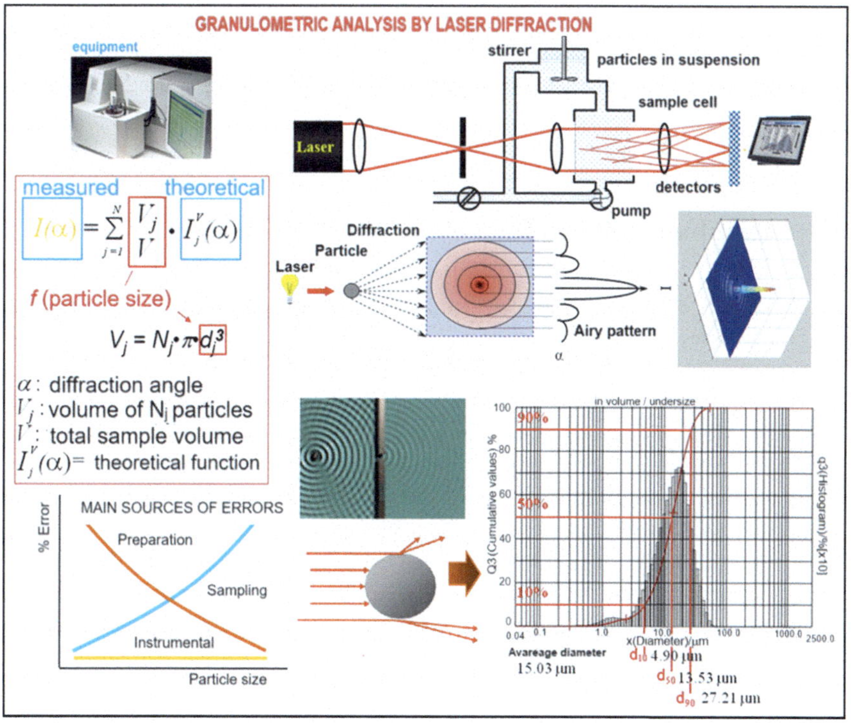

Fig. 3.3 Characteristics of granulometric analysis by LASER diffraction

than the length of the LASER radiation used ($\gg$ 830.625 nm). In addition, the distance between the particles should be much larger than their diameter, and their positioning should be random.

Measurements with a granulometer by LASER diffraction present at least three important advantages. First, the measurement is independent of the spectrum of compositions and the distribution of phase densities in the sample, as long as each particle has an equal probability of diffracting the LASER beam. This can be ensured by using a turbulent and diluted medium, containing sufficiently small particles. A second advantage is the speed of the analysis. A sample can be measured in seconds. The third advantage is the relative ease with which an analysis can be obtained.

Despite these important advantages, the granulometer by LASER diffraction as an instrument for granulometric analysis is limited to relatively small sizes, below 3 mm, while sieves are used in a spectrum of larger sizes, up to several tens of millimeters. As most of the unit operations in ore treatment involve larger particles the range of use of the granulometer by LASER diffraction, sieving meets most of the needs in industrial operations in Mineral Technology. A second major disadvantage of the granulometer by LASER diffraction is that the granulometric fractions after analysis are not physically separated, and chemical and mineralogical analyses, often necessary, based on the granulometric fractions, cannot be immediately performed after the test.

The analysis can be performed wet or dry. In the dry method, the sample needs to have good fluidity. If the sample is in suspension, control over the fluid used, dispersing agent, time of dispersion (ultrasound is occasionally used for greater dispersion).

3.4 Granulometric Analysis by Sedimentation

Granulometric analysis by sedimentation is based on the terminal sedimentation speed of particles in a fluid. Examples include the Andreasen pipette and the sedimentograph, which apply Stokes' law, which is valid only for Reynolds numbers less than 0.2. These techniques should be applied to materials that have only one constituent, since, in the calculation of particle size, the density of the material is used. Therefore, this is a restriction to its use in the analysis of mineral raw materials, notably ores, composed of more than one mineral phase. Figure 3.4 presents some characteristics of the granulometric analysis by sedimentation.

Other disadvantages of granulometric analysis by sedimentation include the slowness of the measurement. Particles of irregular shape, like kaolin, take even more time to settle, due to their geometry that promotes drag, compared to spherical particles. The technique also has a limited range, with specific difficulties below 2 mm and above 50 mm.

Temperatures also require close monitoring to control viscosity. For each variation of 1 °C in temperature will produce a change of 2% in viscosity.

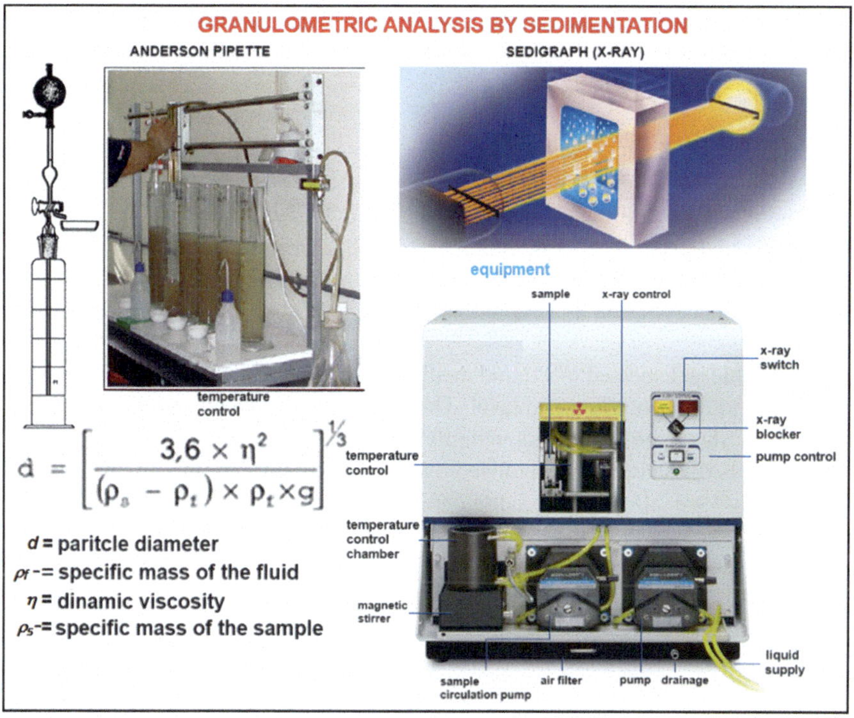

Fig. 3.4 Characteristics of granulometric analysis by sedimentation

Stokes' Law predicts that a particle falling in a vacuum, under the action of gravity, does not suffer resistance to sedimentation. Therefore, the sedimentation speed increases indefinitely, regardless of its size and density. The movement of this same particle, if immersed in any fluid medium, is subject to a resistive force, whose magnitude depends on the prevailing fluid dynamic regime, in addition to the morphological aspects of the particle. When equilibrium is reached between the gravitational force and the fluid's resistance force, the particle reaches its terminal sedimentation speed at a constant rate.

Stokes' law applies only to the region of laminar flow. It is common to admit as the limit of purely laminar flow the value Re = 0.2. Substituting this value into the Reynolds number expression and applying the speed obtained in Stokes' equation, we arrive at the maximum particle diameter determinable (Eq. 3.11).

$$d = \left\{\left(3.6 \times \eta^2\right)/\left(\rho_s - \rho_f\right) \times \rho_f \times g\right)\right\}^{1/3} \tag{3.11}$$

where:

d = particle diameter;

ρ_f = specific mass of the fluid;

η = dynamic viscosity;

ρ_s = specific mass of the sample.

3.4.1 Andreasen Pipette

The Andreasen pipette consists of a graduated cylinder from 0 to 20 cm, whose volume is equal to 0.5 L. This is connected to a 10 ml reservoir through a two-outlet tap. The procedure for drawing up the granulometric curve consists of measuring the variation in concentration at a certain level within the dispersion, at predetermined time intervals.

The sample is dispersed in a liquid medium (usually aqueous) and poured into the cylinder up to the predetermined level. The pipette is introduced and the suspension is stirred by inversion. After this agitation, it is left to settle and at predetermined time intervals the samples are removed by suction at the top of the reservoir. Then, the tap is reversed, allowing the sample to drain to the collector. After the sample is removed, the new level of the liquid in the pipette is noted.

3.4.2 Sedigraph

The sedigraph consists of a beaker, magnetic stirrer, one or two peristaltic pumps, a cell where the particles settle and a source of X-rays, connected to a computer containing a *software* that calculates the terminal sedimentation speeds for particles of different diameters.

The vertical distance and fall time are then selected for each diameter to be measured, so that a particle, falling at a certain speed, will descend a certain distance, in a pre-selected time, providing the time that will be spent for the analysis.

In the first stage of the analyses, the sedigraph measures the transmittance of X-rays through the moving sample suspension. This transmittance is considered as an indicator of the homogeneous mixture of all the particles, that is, it is related to the total concentration of particles present in the sample. Then, the equipment stops the movement of the sample suspension, which allows the particles to settle. The X-ray transmittance is then measured again at different positions in the cell (measured in relation to the top of it) and at pre-selected times based on Stokes' law. The transmittances thus obtained relate to the concentration of particles with the exact diameter corresponding to the distance and fall time at which they were measured. In this way, for each desired diameter, the sedigraph provides a ratio that, multiplied by 100, gives the percentage of particles with the desired diameter.

3.4.3 Granulometric Analysis by Microscopy

Granulometric analysis by microscopy is an excellent technique that allows direct analysis of the particles of a sample. However, it is not appropriate as a quality control or production technique. In addition, as a relatively small number of particles are analyzed, there is the real issue of representative sampling. It is recommended that a minimum of 10,000 images be examined for statistical validation. Furthermore, the preparation of samples for electron microscopy is laborious and time-consuming, and by manual methods fewer particles are examined.

Chapter 4
Surface Area and Porosimetry

4.1 Surface Area

Specific surface area is defined as the ratio between the area and the mass of the particle. It is an important parameter for the characterization of raw materials due to its correlation with chemical reactivity and chemical phenomena such as physical adsorption and sintering, which occur from the surface. Figure 4.1 presents some aspects related to the determination of the surface area of a raw material.

The validity of the relationship between particle size and its area Surface area depends on the sphericity of the particle. In practice, particles have an irregular, non-spherical surface, making it unusual to estimate the surface area from the average size of the particles. Figure 4.2 summarizes some characteristics involving the morphology and determination of the surface area of mineral raw materials.

The most used technique for determining the specific surface area is gas adsorption, which measures the amount of gas that is adsorbed on the internal and external surfaces of a porous material, when the partial pressure of an inert gas varies. The amount of gas adsorbed depends on its relative vapor pressure and is proportional to the total external and internal surface of the material.

$$\frac{1}{W\left(\frac{P}{P_0}\right)^{-1}} = \frac{1}{W_m C} + \frac{C-1}{W_m C}\left(\frac{P}{P_0}\right) \tag{4.1}$$

where:

W = amount of gas adsorbed (at relative pressure P/P_0)

P_0 = ambient pressure;

$P \rightarrow$ pressure during analysis;

C = BET constant;

W_m = mass of an adsorbed monolayer

Figure 4.3 illustrates the relationship between these parameters and the determination of the specific surface area of a material.

C. P. Bergmann, *Techniques for Analysis of Mineral Raw Materials*, Topics in Mining, Metallurgy and Materials Engineering, https://doi.org/10.1007/978-3-031-58985-0_4

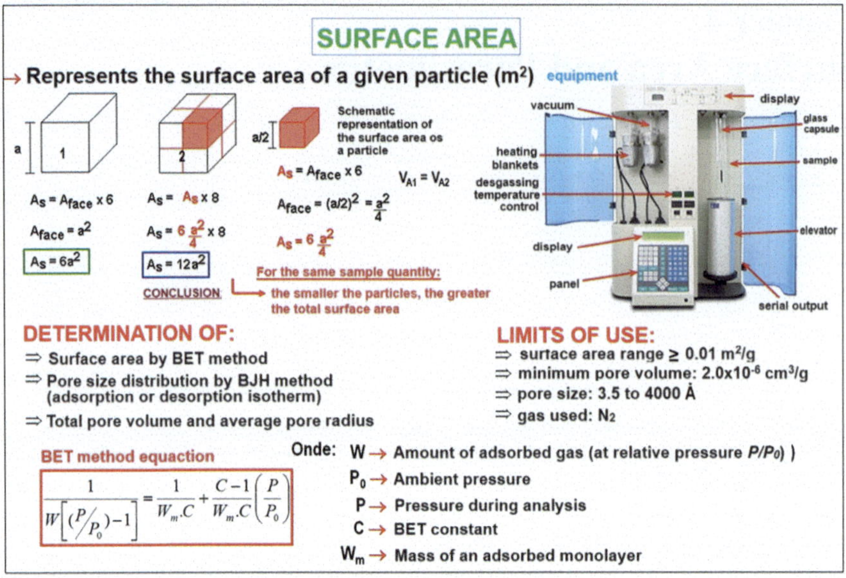

Fig. 4.1 Aspects related to the determination of the surface area of particles in a raw material.

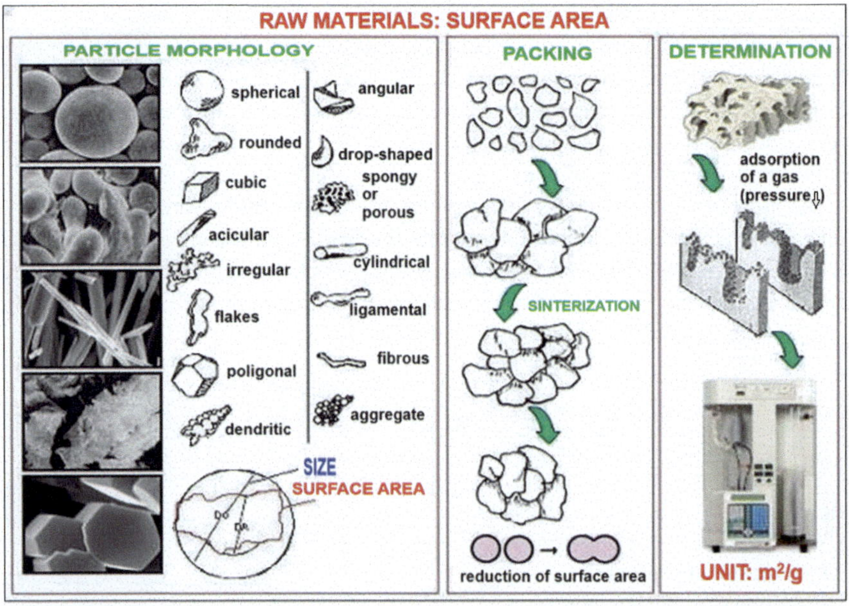

Fig. 4.2 Some characteristics involving the morphology and determination of the surface area of mineral raw materials.

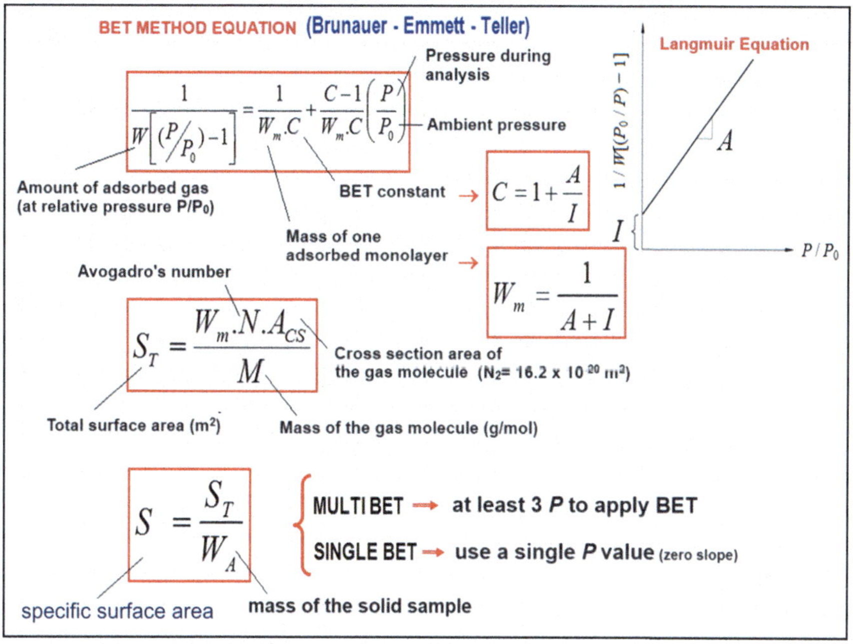

Fig. 4.3 Relationship between the parameters of gas adsorption on the surface of a material and the determination of its specific surface area

The gas used can be nitrogen and helium. The relationship between the amount of gas adsorbed and relative vapor pressure at a constant temperature is called an **adsorption isotherm**. The usual unit is m^2/g. This measure is generally referred to as the **BET** (**Brunauer, Emmett, and Teller**) specific surface area in reference to the authors of the method in 1938.

To minimize errors in the measurement, the sample should be completely dried, and it should not be hydrophobic. The analysis should be performed as quickly as possible to avoid gas desorption.

The determination of the specific surface area by the **BET theory**, developed The current **IUPAC classification** provides for 5 types of isotherms (Fig. 4.4). Curve type I represents the adsorption isotherm of microporous materials. Types II and III describe adsorption in macroporous or non-porous materials with strong and weak adsorbate-adsorbent interactions, respectively. Curve types IV and V are associated with types II and III, respectively, presenting adsorption curves with hysteresis cycle. Curve type VI presents adsorption on non-porous surface in stages, represented by steps. Table 4.1 presents the IUPAC classification of pores according to their size.

The method BET continues to be widely used in studies of surface area and porosity of materials, due to its simplicity and ability to accommodate the five main types of adsorption isotherms (I to V). For its greater validity, the experimental conditions must be as close as possible to the following assumptions:

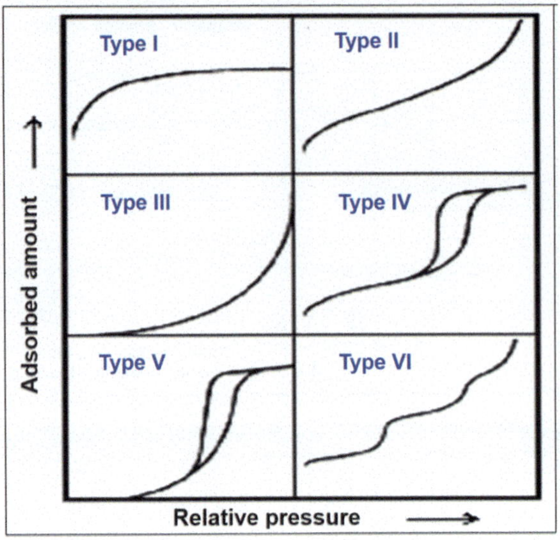

Fig. 4.4 Types of isotherms according to IUPAC

Table 4.1 Classification of pores by size according to IUPAC

Classification	Size range
Micropores	0–2 nm (0–20 Å)
Mesopores	2–50 nm (20–500 Å)
Macropores	50–7500 nm (0.05–7.5 μm)
Megapores	>7500 nm (>7.5 μm)

(i) adsorption is caused by van der Waals forces (physical adsorption);
(ii) the adsorption of the 1st layer is carried out following the Langmuir model, on the surface sites;
(iii) the molecules adsorbed in the first layer constitute sites of adsorption for the formation of the second layer, etc.;
(iv) the number of sites, as well as the size of the surface, are constant for each layer;
(v) all sites of the same layer are energetically equivalent and there are no lateral interactions between the adsorbed molecules;
(vi) the adsorption energy is constant in the first layer;
(vii) the adsorption energy in the other layers is equal to the energy of condensation.

4.2 Porosimetry

The **porosity** is a microstructural characteristic of fundamental importance as it decisively influences several other properties of technological interest.

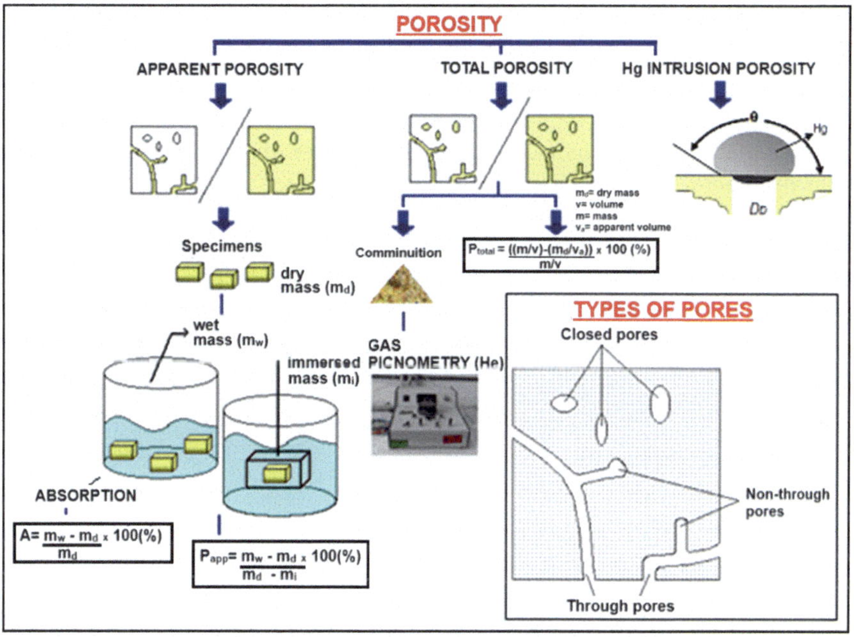

Fig. 4.5 Types and determination of porosity (including water absorption)

Closed and isolated pores make up the porosity. Open pores have a connection with the external surface, that is, they are permeable to fluids, and can be filled by them. Depending on the intercommunication of the pores, porosity can be classified as closed or open, and this as transpassing and non-transpassing (Fig. 4.5).

Apparent porosity is defined as the ratio of the volume of open pores to the apparent volume of the sample, expressed as a percentage. While total porosity is the ratio of the total volume of open and closed pores to the apparent volume.

4.2.1 Determination of Porosity by Water Absorption

The determination of apparent porosity is commonly done by the Archimedes method. Total porosity can be calculated by the difference between real and apparent densities, according to Eq. 4.2.

$$P_{total}(\%) = \frac{\left(\left(\frac{m}{v}\right) - \left(\frac{ms}{v.a}\right)\right)}{\frac{m}{v}} \times 100 \tag{4.2}$$

where:

P_{total} = total porosity;

m = sample mass;
m_s = dry mass of the sample;
$v.a.$ = apparent volume of the sample;
v = sample volume.

Porosity can also be expressed in the form of **water absorption** which is linked to the open porosity of the material, and can be measured by the ratio between dry weight and wet weight. Normally, water is used, which has established the test as **water absorption**, but as many raw materials react with water, it is common to use kerosene as the saturating liquid. Absorption is calculated by Eq. 4.3.

$$A = \frac{(mu - ms)}{ms} \times 100 \tag{4.3}$$

where:
A = water absorption (%);
m_u = wet mass (g);
m_s = dry mass (g).

4.2.2 *Determination of Porosity by Gas Expansion*

For the determination of porosity by the gas expansion method, also called **pycnometry**, a sample of known volume is placed in a container of known volume, connected to another container also with known volume, under vacuum (Fig. 4.6). When a valve connecting the two containers is opened, the gas passes from the first container to the second, until a uniform pressure distribution is achieved. Using the ideal gas law, the volume of the pores is calculated by Eq. 4.4.

$$V_V = V_T - V_a - V_b . \left(\frac{P_2}{(P_2 - P_1)} \right) \tag{4.4}$$

where
V_V = effective volume of pores;
V_T = total volume of the sample;
V_a = volume of the container that contains the sample;
V_b = volume of the container under vacuum;
P_1 = initial pressure in V_a and V_b;
P_2 = final pressure throughout the system.
The porosity φ is obtained directly by Eq. 4.5.

$$\Phi = \frac{V_V}{V_T} \tag{4.5}$$

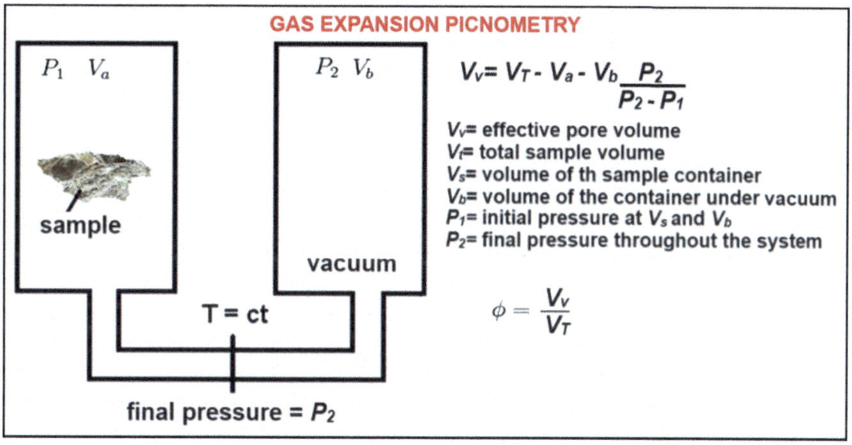

Fig. 4.6 Determination of porosimetry by gas expansion

This method assumes that the gas reaches all pores. For this reason, the smaller the size of the gas (atom or molecule) used, the greater will be the probability of satisfying this condition.

4.2.3 Determination of Porosimetry by Mercury

Porosimetry is an analytical technique used to determine various quantifiable aspects of a porous material porous. The technique involves the intrusion of a low-wettability fluid (usually mercury) at high pressure into a porous sample through the use of a **porosimeter**. The size of the pores can be determined based on the external pressure required to force the fluid to enter the pores against the opposing force of the fluid's surface tension.

Mercury porosimetry is extremely important for the determination of material porosity, as in a single measurement of duration of approximately 30 to 45 min it is possible to determine the apparent density, the solid density, the size distribution of pores from the total pore volume and the specific area of the material.

Mercury is most often used in the determination of porosity, as it does not wet most materials and is non-reactive to most materials at room temperature.

The contact angle (or wetting angle) measures the wettability between a liquid and a surface. It is determined by the relationship between the energies of the involved interfacial surfaces: solid-liquid (γ_{SL}), solid-vapor (γ_{SV}) and liquid-vapor (γ_{LV}). The **Young's Equation** establishes the relationship between the involved surface energies (Eq. 4.6).

$$\gamma SV = \gamma SL + \gamma LV . \cos\theta \qquad (4.6)$$

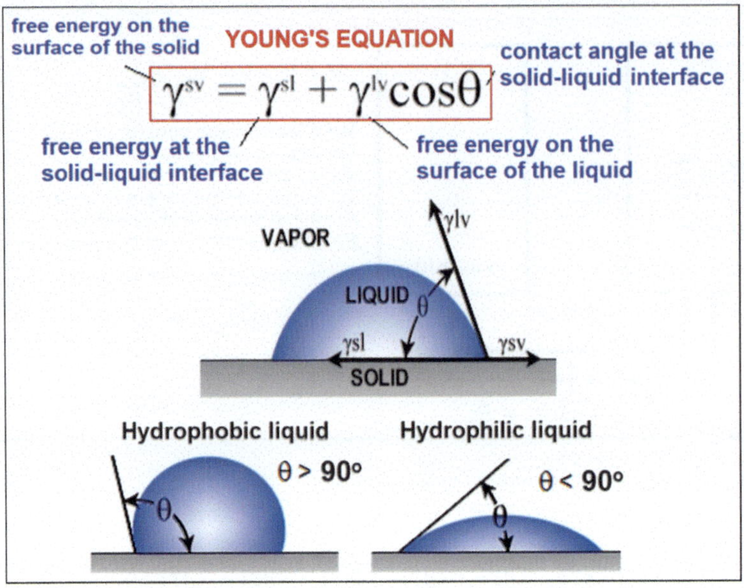

Fig. 4.7 Equilibrium situation in the wetting of a solid by a liquid: relationship between the involved interfacial energies and the contact angle (or wetting angle)

where:

γ_{SV} = free energy at the solid-vapor interface

γ_{SL} = free energy at the solid-liquid interface

γ_{LV} = free energy at the liquid-vapor interface

When the contact angle is greater than 90°, then the liquid does not wet the solid, that is, there is no formation of an extensive solid-liquid surface, and the liquid remains in the form of a drop. Figure 4.7 illustrates the equilibrium situation in the wetting of a solid by a liquid.

The force balance equation known as Washburn's Equation for the described situation, assuming the pores with cylindrical shape is given by Eq. 4.7.

$$P_L - P_G = \frac{(4\sigma.\cos\theta)}{D_p} \tag{4.7}$$

where:

P_L = liquid pressure;

P_G = gas pressure;

σ = liquid surface tension;

θ = contact angle of the intruding liquid;

D_P = pore diameter.

As the technique is usually carried out under vacuum, the gas pressure starts at zero. The contact angle of mercury with most solids is between 135° and 142°, so

~140° can be taken as the average value. The surface tension of mercury at 20 °C under vacuum is 480 mN/m. Thus, for a large number of substances, Eq. 4.8 is valid.

$$Dp = \frac{(1470\,\text{kPa}\,.\mu\text{ m})}{P_L} \tag{4.8}$$

Figure 4.8 illustrates aspects of the determination of porosimetry by mercury intrusion. As the pressure increases and there are no pores of size that satisfy Eq. 4.3, there is no mercury intrusion and the intrusion volume remains equal to zero (region I). When the pressure is sufficient to fill pores of a certain size, there is mercury penetration and variation of intrusion volume (region II). After the pores are filled with mercury and there are no more pores that satisfy Eq. 4.7, the penetration of mercury ceases and no longer penetrates the sample (region III).

The maximum value of intrusion volume, V_M, determines the volume of open pores of the sample, while the pore size for $V_M/2$ can be considered the average pore size. The representation of the curve can be made as a function of the volume variation for a pressure increment.

The determination of the **surface area by mercury intrusion**, should consider the reversible work dW required to wet an area dA of a pore surface not wetted by mercury, given by Eq. 4.9.

$$dW = \gamma \cdot \cos\phi \cdot dA \tag{4.9}$$

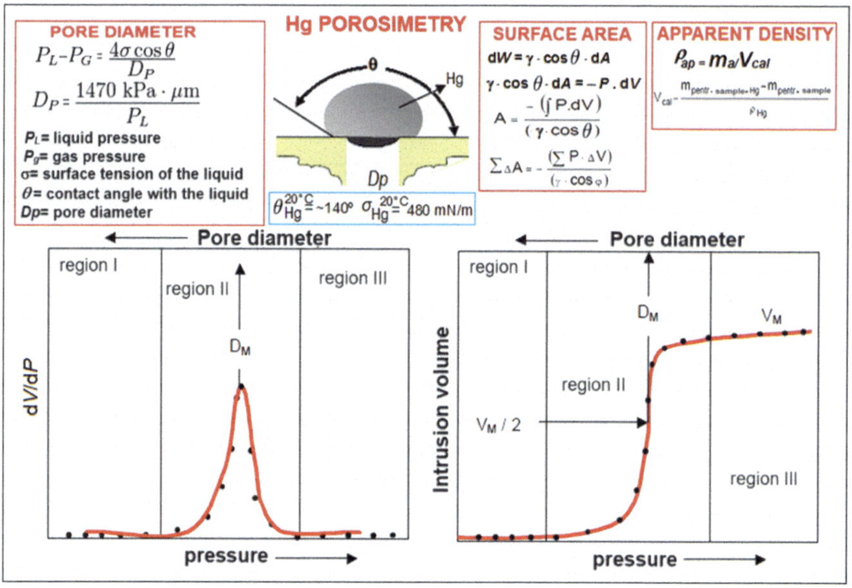

Fig. 4.8 Mercury in contact with a porous solid where θ is the contact angle and D is the pore diameter. Typical curves of a mercury porosimetry analysis

where γ is the surface tension of mercury and ϕ is the contact angle with the pore surface. In the case of mercury penetrating the pores, the work is provided when the external pressure P forces a volume of mercury d V into the pores, according to Eq. 4.10

$$\gamma \cdot \cos\phi \cdot dA = -P.dV \tag{4.10}$$

Assuming that γ and ϕ do not vary with pressure, Equation 4.10 can be rewritten in the form of Eq. 4.11.

$$\mathrm{A} = -\frac{\left(\int P.\mathrm{d}V\right)}{\gamma \cdot \cos\phi} \tag{4.11}$$

In this way, through the pressure increments and their respective variations of pore volume, it is possible to determine the increments of area. The sum of these increments, given by Eq. 4.12, is the total specific area in m^2/g.

$$\Sigma\Delta A = -\frac{(\Sigma P.\Delta V)}{\gamma \cdot \cos\phi} \tag{4.12}$$

For the **determination of apparent density** during mercury porosimetry, the volume of mercury displaced by the sample is considered. Equation 4.13 allows calculating this volume, and considering the mass of the sample, its apparent density is obtained, according to Eq. 4.14.

$$V_{calc} = \frac{\left(m_{fullHg} - m_{sample}\right)}{\rho_{Hg}} \tag{4.13}$$

$$\rho_{app} = \frac{m_{sample}}{V_{calc} - \left\{\frac{\left(m_{pentr+sample+Hg} - m_{pentr+sample}\right)}{\rho_{Hg}}\right\}} \tag{4.14}$$

where:

V_{calc} = volume of displaced mercury
ρ_{app} = apparent density
m_{sample} = sample mass
m_{fullHg} = mass of the sample fully filled with Hg
$m_{pentr+sample+Hg}$ = mass of penetrated mercury and mass of the sample fully filled with Hg
$m_{pentr+sample}$ = mass of penetrated mercury and mass of the sample
ρ_{Hg} = Hg density.

Chapter 5
Thermal Analysis

5.1 Introduction

Thermal analysis (or generically, thermoanalysis) involves a series of techniques that evaluate the transformations that occur in a substance, as a function of temperature. Although it has been used for a long time as a research tool in the laboratory, more recently, it has been introduced in industrial practices such as, for example, in process control and inspection of raw materials and materials.

For your thermal analysis, the sample is placed in a device in which it is possible to detect based on temperature and/or time physical or chemical changes in its structure, producing an electrical signal analogous to the observed change. This signal is electronically amplified and transferred to a reading device in a recorder. Figure 5.1 presents the main techniques of thermal analysis, highlighting those applied to mineral raw materials.

In the analysis of mineral raw materials, basically the techniques of thermogravimetric analysis (TGA) and differential thermal analysis (DTA) are used, and, less frequently, a similar one to this, differential scanning calorimetry (DSC).

In TGA, the mass of the sample is monitored as a function of temperature, through a microbalance, which records and quantifies any loss or gain of mass. In the case of differential thermal analysis (DTA), a thermocouple analyzes the sample temperature and compares it with the temperature of an inert substance, in the same furnace. The equipment used usually contemplate DTA/TGA simultaneously.

In addition to the mentioned thermal analysis techniques, one can also submit the gases possibly dispensed to analysis (GA), including chemical or characteristics such as pressure of volatile substances in continuous vacuum systems, recording curves with volatilization speed as a function of temperature (or time). Thermal analysis systems can be connected to **chromatographs, mass spectrometers or infrared**, for the analysis of emanated gases.

The equipment used in thermal analysis is basically constituted by a furnace, thermocouples, microbalance and a gas flow system. The sample size varies from a

C. P. Bergmann, *Techniques for Analysis of Mineral Raw Materials*, Topics in Mining, Metallurgy and Materials Engineering, https://doi.org/10.1007/978-3-031-58985-0_5

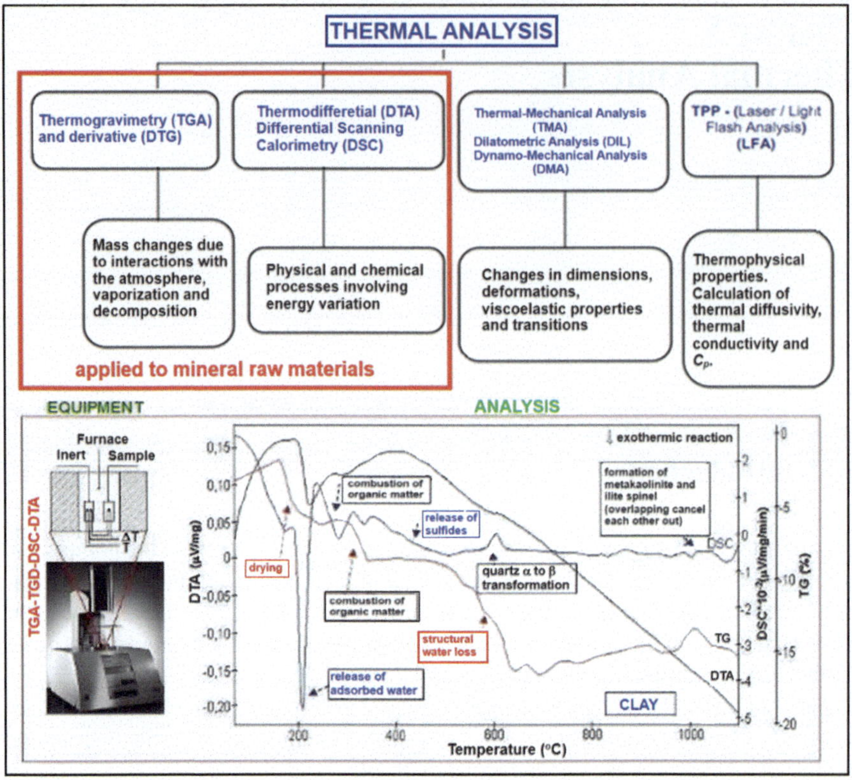

Fig. 5.1 Main techniques of thermal analysis, highlighting those applied to mineral raw materials

few mg to 10 mg depending on the equipment used. Nitrogen (N_2) and argon (Ar) are commonly used to purge the furnace and prevent the oxidation of the sample. Figure 5.2 presents a schematic drawing of a thermal analysis equipment.

The specific details of each equipment depend on the application and the manufacturer. There are four types of furnaces used: from −150 to 500 °C, 25 to 1000 °C, 25 to 1600 °C, and 25 to 2800 °C. The heating rate, in general, varies from 1 to 50ºC/min. A typical sensitivity of 0.1μg is possible, but not easily achieved under temperature change conditions (more realistically sensitivity of 1μg). The type and nature of the crucible, which can be platinum, alumina, quartz or glass, depends on the maximum exposure temperature, the chemical nature of the sample, its quantity and reactivity.

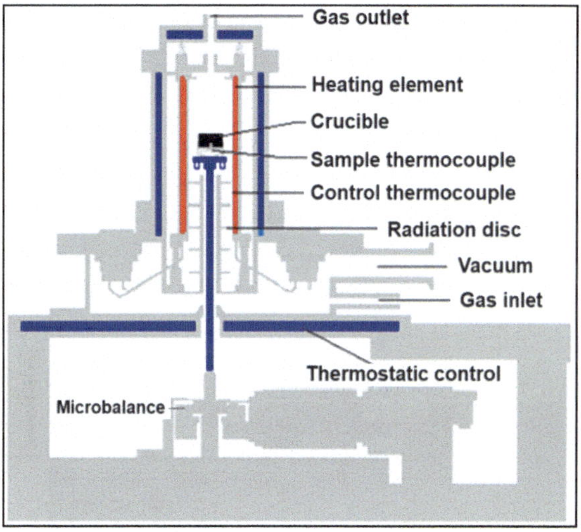

Fig. 5.2 Representation of a thermal analysis equipment and its main parts

5.2 Thermogravimetric Analysis (TGA)

The **thermogravimetric analysis** (TGA) or **thermogravimetry** (TG) is the technique used to measure the mass variations of a sample as a function of temperature. The main phenomena measured in this technique include: evaporation, sublimation, decomposition, oxidation, reduction, adsorption and desorption of gases.

The thermogravimetric curve can be interpreted by its derivative in relation to time, called the DTG (differential thermogravimetry) curve, as shown in Fig. 5.3.

Figure 5.4 presents different types of typical curves in a thermogravimetric analysis, described below:

(i) characteristic of a sample that does not present mass variations in the applied thermal cycle;
(ii) the rapid initial loss of mass is in many cases associated with the loss of sample moisture or the desorption of gases;
(iii) represents the decomposition of a sample in a single stage;
(iv) presents multiple stages of decomposition, where it is possible to determine the stability limit temperatures of the reactants.
(v) the same as curve iv, but the sample was subjected to lower heating (or cooling) rates;
(vi) represents mass gain, as occurs for example during the oxidation of metals.
(vii) rarely observed. An example is the oxidation of silver and subsequent decomposition, at high temperatures, of the formed oxide.

The stages present in the curves of type (vi) and (v) of Fig. 5.4 can be clearly identified as peaks in the DTG curves, as shown in Fig. 5.5.

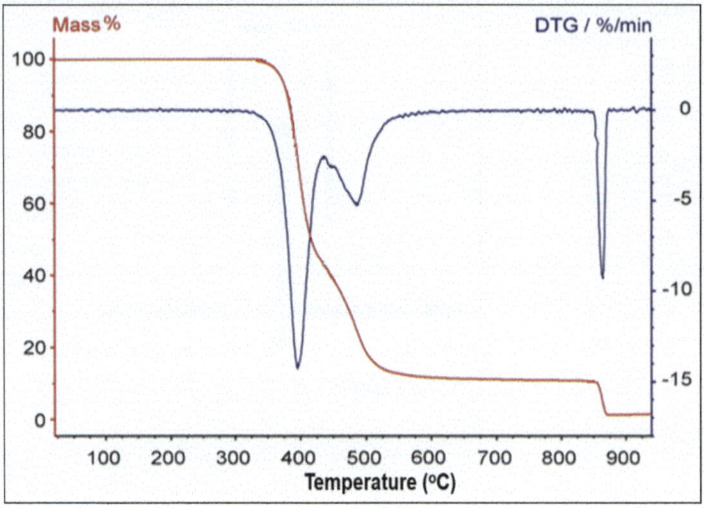

Fig. 5.3 Typical curves of a thermogravimetric analysis, representing the TGA and its DTG

Fig. 5.4 Main types of typical thermogravimetric curves (TGA)

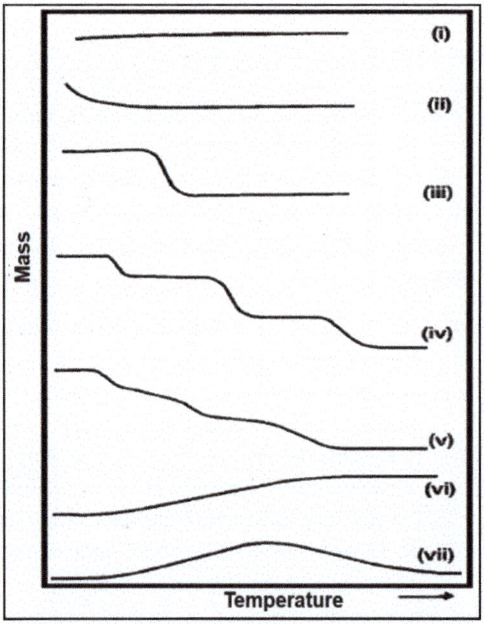

5.2.1 *Applications of TGA*

Among the applications of thermogravimetric analysis in the characterization of mineral raw materials, the following stand out:

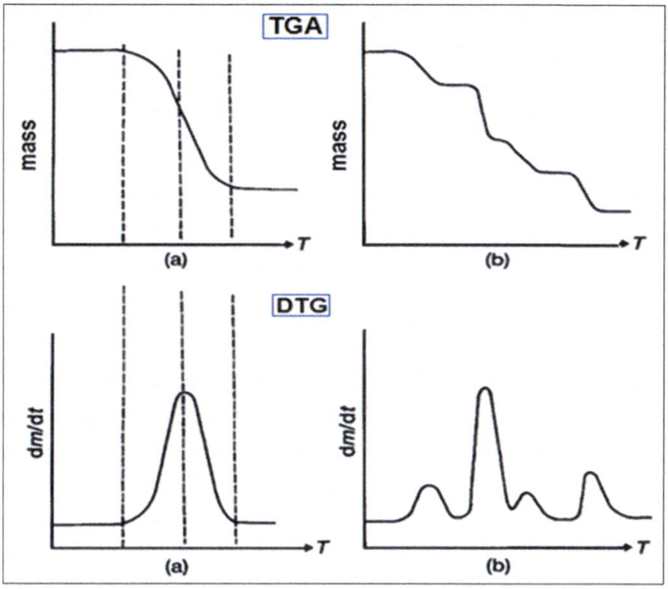

Fig. 5.5 Comparison between TGA and DTG curves

(i) calcination of ores; (ii) adsorption and desorption curves; (iii) thermal decomposition or pyrolysis of organic and inorganic materials; (iv) determination of vapor pressure and enthalpy of vaporization of volatile additives; (v) determination of moisture, volatile species and evaporation and sublimation rates; (vi) study of dehydration and hygroscopicity; (vii) magnetic properties such as Curie temperature, magnetic susceptibility.

5.3 Differential Thermal Analysis (DTA)

The **Differential Thermal Analysis (DTA)** measures the temperature difference between the sample and an inert standard (usually alumina), used as a reference, as a function of temperature. While the standard substance linearly increases its temperature with the supply of heat, temperature changes in the sample can be caused by enthalpic transitions (exothermic or endothermic). Generally, phase transitions, dehydrations, reductions, and some decomposition reactions produce endothermic effects, while crystallization, oxidation, and other decomposition reactions produce endothermic effects.

Figure 5.6 shows the diagram of the DTA compartment in which two crucibles (of the sample to be analyzed and the reference material), and two temperature sensors (one sensor in each crucible), in a system heated by only one heat source.

Fig. 5.6 Schematic diagram of the sample compartment in DTA analysis

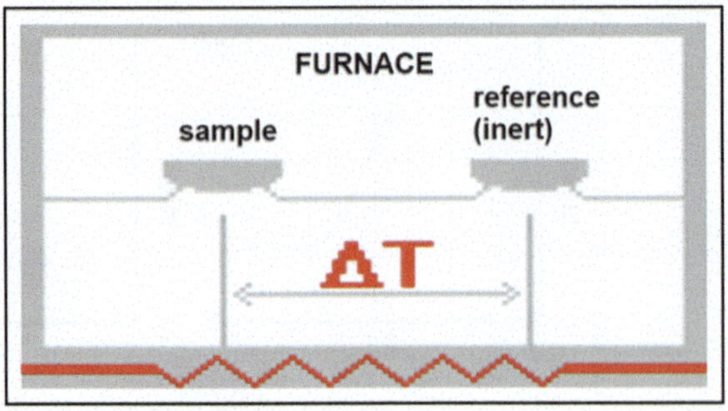

When reactions are exothermic, the heat produced in the sample reaction heats the sample more intensely than the furnace heats the inert material. For endothermic reactions, the contribution of external heat is spent maintaining the reaction temperature, causing the sample temperature to be lower than the inert material. The equipment records the temperature differences of the sample and the inert material in the form of a **thermal or thermodifferential curve**. The exo- or endothermic reactions appear on the curve as inflections.

The DTA curve is then recorded with temperature or time on the abscissa and μV on the ordinate. The temperature difference is given in μV due to the use of thermocouples in its measurement (Fig. 5.7).

The main use of DTA is to detect the temperature of thermal processes and qualitatively characterize them as endothermic and exothermic, reversible or irreversible, first order transition or second order, etc. This type of information, as well as its

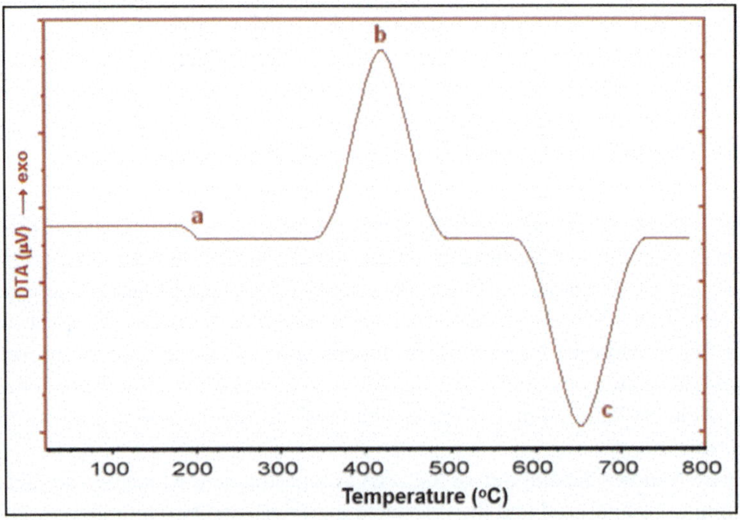

Fig. 5.7 Typical curve of differential thermal analysis (DTA): **a** variation of heat capacity, **b** exothermic reaction, **c** endothermic reaction

dependence on a specific atmosphere, makes this method particularly valuable in determining phase diagrams.

5.3.1 Differential Scanning Calorimetry (DSC)

Differential scanning calorimetry (DSC) is a technique derivative of TDA. The DSC was developed with the intention of quantifying the energy involved in reactions. Figure 5.8 presents thermal analysis DSC of two originally hydrated inorganic materials. The area of the peak directly under the mW/mg curve is proportional to the enthalpy ΔH involved in the endothermic/exothermic process, expressed in J/g or kJ/kg.

There are two types of equipment that perform the Differential Scanning Calorimetry test (Fig. 5.9):

(i) *DSC* energy compensation: the sample and the reference have individual heating sources. The temperature and energy are monitored and generated by identical platinum filaments, thus acting as resistive thermometers and heaters, keeping the heat provided constant. However, instead of measuring the temperature difference between the sample and the reference during the reaction, a control system increases immediately the energy provided to the sample when the process is endothermic, and increases the energy provided to the reference

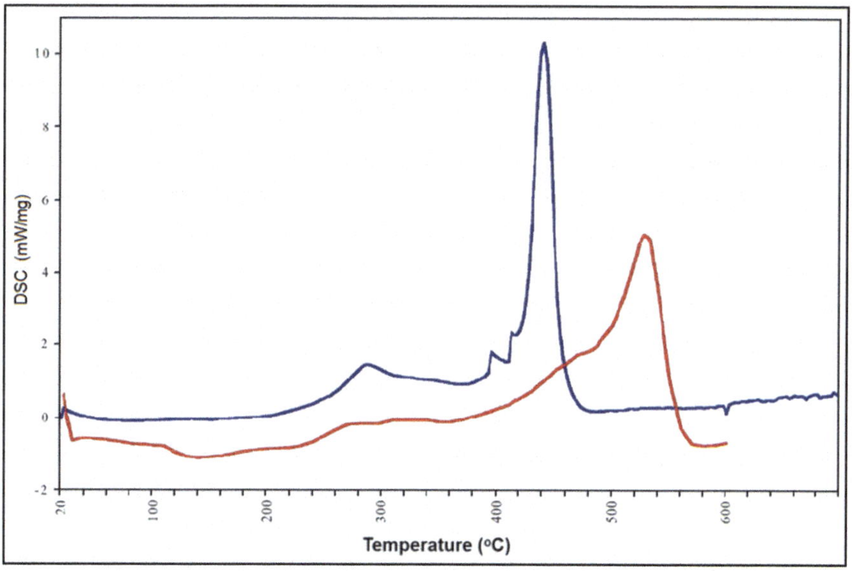

Fig. 5.8 Thermal analysis *DSC* for two distinct originally hydrated inorganic materials

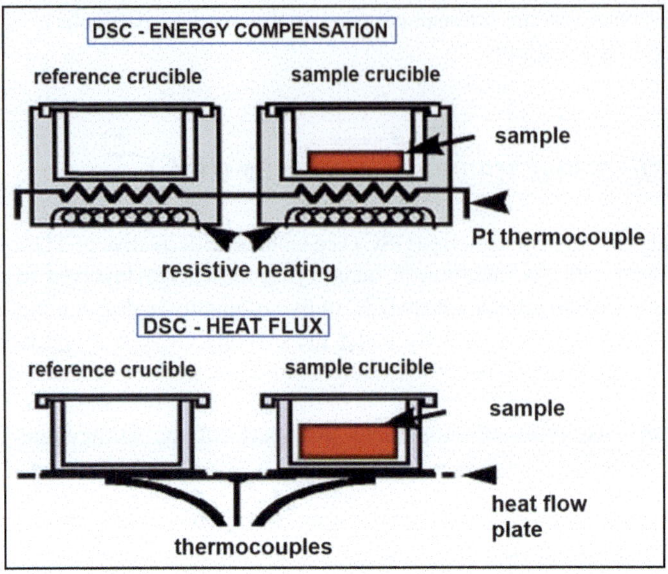

Fig. 5.9 Types of equipment that perform the Differential Scanning Calorimetry test

when the process is exothermic, thus keeping the sample and the reference at the same temperature.

(ii) *DSC* by heat flow: quite similar to DTA, since only one furnace is used. The crucibles are arranged on a base of a highly conductive metal, usually platinum. The sample and the reference are then heated by the same energy supply system. Each time the sample reacts, an energy flow is established between the crucibles through the platinum base. The heat flow is then measured through the temperature sensors positioned under each crucible, thus obtaining a signal proportional to the difference in thermal capacity between the sample and the reference.

Heat compensation equipment is generally limited to temperatures up to 725 °C, while heat flow ones can operate up to 1500ºC. The results of the *DSC* curve, in both techniques, are expressed in terms of temperature or time on the ordinate, and mW/mg (milliwatts per milligrams) on the abscissa.

5.3.2 Applications of TDA and DSC

Among the applications of the **TDA** and **DSC** analysis in the characterization of raw mineral materials, the following stand out:

(i) heat capacity; (ii) determination of purity; (iii) phase diagrams assembly; (iv) enthalpy of transitions; (v) thermal and oxidative stability; (vi) degree of crystallinity; (vii) melting range; (viii) nucleation; ix) crystallization and reaction rates.

5.4 Influence Factors and Sources of Error

The factors that can influence thermal analysis are of two types: instrumental factors (oven heating rate, oven atmosphere, sample holder and oven geometry) and factors related to the characteristics of the sample (particle size, quantity, compaction, thermal conductivity and nature of the sample, solubility of the gases released in the sample itself, heat from any chemical/physical reactions). Some causes of error are well known, among these are highlighted.

(i) air thrust on the crucible and its support. This effect causes an apparent increasing weight of the crucible as the temperature is raised, and can be determined by obtaining a thermogravimetric curve of the empty crucible. This apparent increase in weight also depends on the material of the crucible.
(ii) convection currents and furnace turbulence. The weight loss caused by the upward flow of hot air under the crucible and the weight gain due to air turbulence depend directly on the size and shape of the crucible.
(iii) temperature measurements and calibration. Normally, it is considered that the sample temperature is the temperature indicated by the thermocouple located in its vicinity. In reality, the sample temperature can be slightly higher or lower than the temperature indicated by the thermocouple, depending on the type of decomposition reaction that is occurring (exo- or endothermic), the geometry of the crucible and support, heating rate used and the thermal conductivity of the sample.
(iv) other errors: fluctuations in the balance system; induction effects caused by the furnace; electrostatic effects on the parts that make up the balance suspension; reactions of the sample with the crucible.

The possibility of error can be minimized by obtaining a baseline, performing the test with two empty crucibles, where these effects are subtracted, trying to obtain a line as horizontal and as close as possible to zero value. This would be expected in the sense that there is no reaction occurring and, therefore, there should not be voltage difference between the two crucibles. Baseline values, to be considered reproducible, should have a difference equal to or less than 0.50 μV for the same temperature.

5.5 Interpretation of Thermal Analysis Data

Physical changes in mineral raw materials are related with the loss of water (humidity), polymorphic transitions (example: quartz-α to quartz-ß) and melting. Chemical processes are more diverse, such as for example:

(i) dissociation: $Ca(CO_3) \rightarrow CaO + CO_2$;
(ii) dehydration: $Ca(SO_4)0.2H_2O \rightarrow Ca(SO_4) + 2H_2O$;
(iii) decomposition: $Fe(CO_3) \rightarrow FeO + CO_2$;
(iv) decomposition and exchange: $Ca(CO_3) + SiO_2 \rightarrow Ca(SiO_3) + CO_2$.

The same processes in different minerals are characterized by different temperatures, due to differences in chemical and mineralogical composition. Chemical impurities also affect temperature changes in mineral raw materials.

Many of the thermal reactions are accompanied by weight change in the sample: drying, dehydration, and loss of volatile components such as CO_2 and SO_2. Oxidation can also result in an increase in sample mass. Some reactions produce heat (exothermic reactions) and occur spontaneously, after reaching the reaction temperature. In other cases, the heated sample needs an additional heat supply to initiate and develop the reaction (endothermic reactions).

The basic reactions in mineral raw materials serve as a basis for the interpretation of their thermal analysis curves. For example, one can cite:

(i) **Native metals**: can undergo oxidation (exothermic) and fusion (endothermic);
(ii) **Sulfides**: when heated they become oxides. Exothermic effects are related to the oxidation of sulfur and its temperature depends on the strength of the bonds between the metals and the sulfur in the initial mineral.
(iii) **Oxides**: can have polymorphic changes during heating (example: quartz, hematite). The oxidation of iron and manganese produces exothermic peaks. Hydroxides have distinct peaks of endothermic dehydration reactions.
(iv) **Carbonates**: their dissociation occurs endothermically. The dissociation temperature depends on the chemical nature of the ore and the bonds present. The presence of iron and magnesium is reflected in the exothermic peak immediately after the dissociation reactions.
(v) **Sulfates**: their heating is characterized by thermal reactions with loss of water and OH^- groups. In some cases, oxidation processes are observed.
(vi) **Silicates**: can show reactions of loss of water and OH^- groups. Exothermic reactions are of oxidation and, in most cases, the formation of new compositions. For example: kaolinite into mullite, forsterite and enstatite by serpentine, among others.
(vii) **Water loss**: Table 5.1 summarizes the type, the form in which it is found in the mineral and the temperature range referring to water present in the mineral.

The thermal analysis of mineral raw materials is widely used in the investigation of structural water loss in hydrated compounds such as clays and in calcination and roasting processes. These analyses are generally conducted in an inert environment

Table 5.1 Type, form in which it is found and temperature range for the reaction, referring to water present in the mineral

Types of water in minerals			
Type of water	Form in which it is found	Dehydration Temperature (°C)	Comments
Absorbed	Non-structural	Up to 100	Water content in minerals depends on the humidity of the environment
Crystallization	Non-structural	100–450	With the loss of mineral water, they can be destroyed (gypsum) or stored (zeolites). Dehydration can be reversible
Constitutional	The OH^- structural group	660–1000	Irreversible loss of water by calcination

(with nitrogen flow being the standard), to prevent the oxidation of some elements (Mn and Fe, for example) from altering the mass variation that will be attributed to the respective minerals.

In the case of coal, on the other hand, synthetic air or oxygen can be used to investigate combustion and the consequent loss of mass by CO_2. In the decomposition of sulfides, on the contrary, the mass can increase by the oxidation of sulfide to sulfate, and decrease by the formation of SO_2 and/or SO_3 (gases eliminated).

Figure 5.10 presents DTA curves of different minerals, containing typical reactions as a function of temperature.

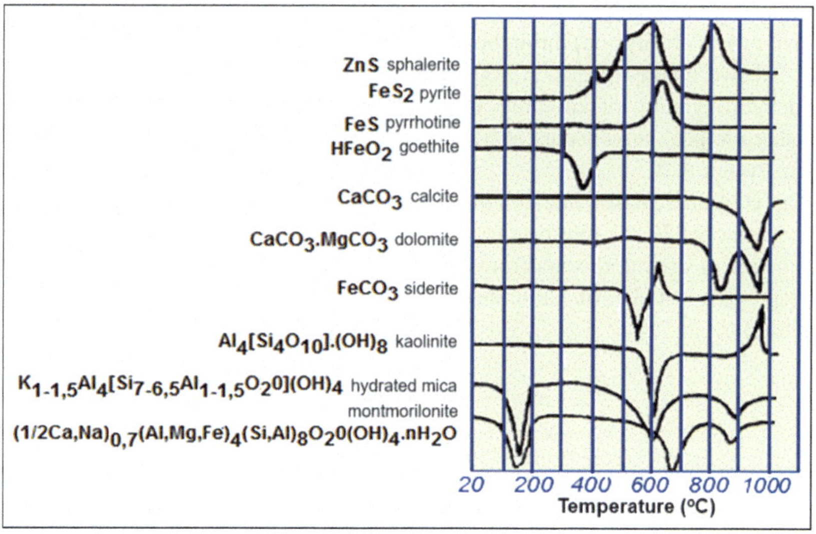

Fig. 5.10 DTA curves of different minerals, containing typical reactions as a function of temperature

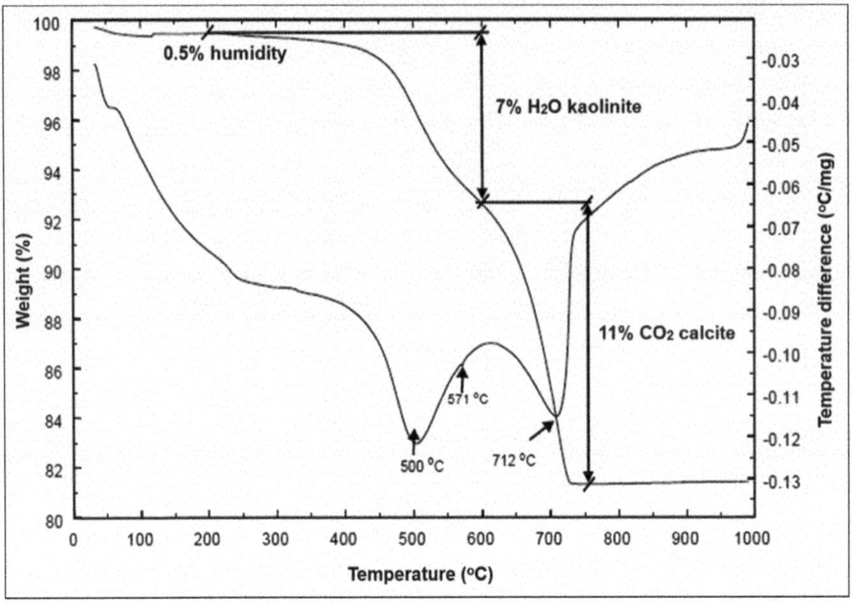

Fig. 5.11 Thermal analysis (thermodifferential in dashed line, and thermogravimetric in solid line), with indication of the endothermic peaks and mass losses, of a sample containing kaolinite, calcite, and quartz

Figure 5.11 presents the TGA and DTA curves obtained for a sample containing kaolinite, calcite, and quartz. The DTA curve (dashed line) indicates an endothermic behavior (heat absorption) throughout the analysis, with the difference always negative, and two marked peaks, at 500.5 and 712.2 °C, corresponding respectively to the dehydroxylation of kaolinite and the decarbonation of calcite. A very small endothermic peak also occurs at 571.5 °C, corresponding to the transformation of quartz from α to β.

The TGA curve (solid line), on the other hand, indicates a small loss of mass up to approximately 220 °C, referring to the sample's humidity. From 220 to 600 °C, the elimination of structural water from kaolinite occurs, and the 7.04% mass eliminated corresponds to 50.3% of kaolinite in the sample (considering 14.0% of H_2O in kaolinite).

Chapter 6
Microscopy Analysis

6.1 Introduction

In the study of mineral raw materials, three main types of microscopy are used: optical microscopy (OM), scanning electron microscopy (SEM) and transmission electron microscopy (TEM). These techniques are complementary and each has its specific field of application:

(i) OM allows the analysis of a larger field of interest, being simple, fast, and inexpensive to use;
(ii) SEM allows in-depth analysis with large magnifications;
(iii) TEM allows analysis with even greater depth of focus;

Table 6.1 summarizes some characteristics of the main types of microscopy. Figure 6.1 schematically presents the operating principles of the main types of microscopes.

6.2 Optical Microscopy

In the characterization of mineral raw materials, the main applications of optical microscopy are: (i) identification and quantification of mineral phases; (ii) composition of mineral phases; (iii) forms of intergrowth and mineral associations.

Transmitted light optical microscopies, for transparent minerals, and reflected light, for opaque minerals, are probably the most traditional methods of mineral identification. Both are based on the interaction of light (usually white light from the visible spectrum) with minerals, and are quite accurate and flexible for qualitative analyses. The analyses require special preparation of the samples, in thin sections (transmitted light), polished (reflected light), or polished thin (both). It is possible

C. P. Bergmann, *Techniques for Analysis of Mineral Raw Materials*, Topics in Mining, Metallurgy and Materials Engineering, https://doi.org/10.1007/978-3-031-58985-0_6

Table 6.1 Some characteristics of the main types of microscopy used in the analysis of mineral raw materials

Characteristic	OM	SEM	TEM
Acceleration voltage (kV)	–	3–50	50–1000
Useful range of magnifications (x)	1–3000	10–50,000	1000–1,500,000
Resolution (Å)	3000	30	3
Depth of focus at 1000 X (μm)	0.1	100	10

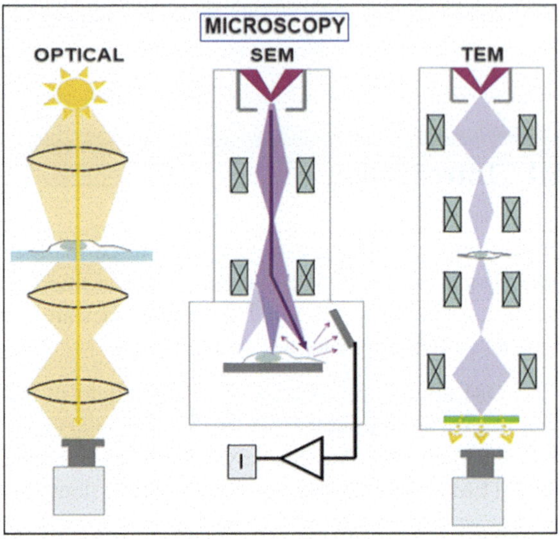

Fig. 6.1 Comparison between the most common types of microscopes

to analyze rock fragments or assemblies of particulate material, embedded in resins (usually epoxy, acrylic or polyester).

A simple optical microscope in its constitution is formed by three elements: an optical magnification system, a light source, and a viewing stage. Figure 6.2 illustrates the elements that typically make up an optical microscope used for petrographic analysis.

Working with a magnifying glass or stereoscopic microscope allows for the analysis of samples without the need to mount polished or thin sections. Minerals are identified by color, shine, habit, cleavages, fracture, and it is possible to use direct auxiliary techniques, such as tweezers to delaminate micas and vermiculite, interaction with a magnet, or specific chemical micro-attacks. For the latter, some grains are placed on Petri dishes or porcelain touch plates.

Analysis with a magnifying glass or stereoscopic microscope is practically standard in a preliminary analysis of a sample, for the identification of more common minerals, evaluation of crystal size, and visual estimation of liberation grain size. The stereoscopic microscope is also essential in the identification, even if qualitative, of

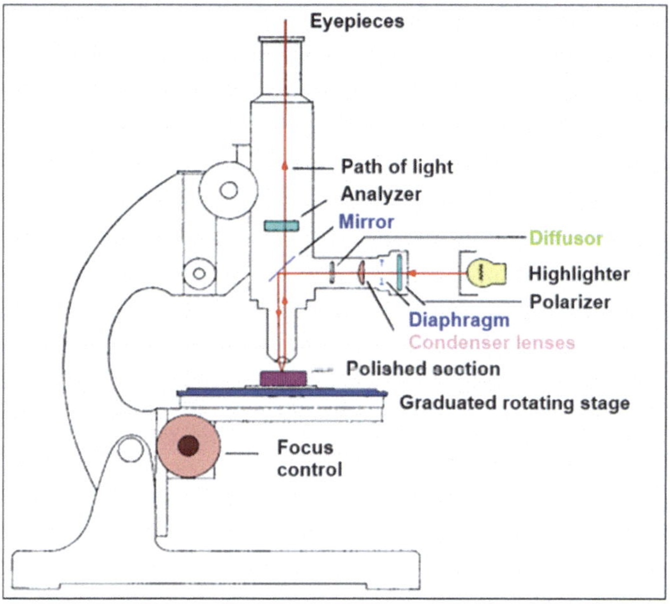

Fig. 6.2 Optical microscope and its main elements, having polarized light for petrographic analysis

forms of surface alteration of grains not visible by more sophisticated techniques, such as scanning electron microscopy.

The quantification of phases in magnifying glass and stereoscopic microscope is possible, and occurs in two ways: visual estimation and picking of phases with subsequent weighing. Visual estimation generally does not present good precision, but often it is sufficient. The assistance of image analysis software significantly increases the precision of the analysis. Despite being laborious, manual picking of phases can be very precise when a statistically significant amount of particles is manually separated.

6.2.1 Petrographic Analysis

The analysis in a petrographic microscope consists of the detailed identification of the rock constituents and is complementary to the analysis, made with hand samples, or to the stereomicroscope (or magnifying glass), and to the examination by optical microscope, by transmitted light. For rocks containing non-transparent metallic minerals (case of several sulfides and oxides, such as pyrite, chalcopyrite, magnetite, ilmenite, etc.), the microscopic examination is performed under reflected light, from extremely well-polished samples, by diamond-based abrasives. Selective mineral staining tests are also used as auxiliary techniques, especially in coarse-grained rocks, in order to highlight structural features, and better evaluate their mineralogical composition.

The Brazilian standard NBR 12768 provides procedures for petrographic analysis of rocks for coatings. The sample should be placed on 2 mm thick glass slides and 0.2 mm thick glass coverslips. For fixation, natural resin glue (Canada balsam) or transparent artificial glue (based on epoxy) and solvent for cleaning, such as xylene or similar, should be used.

The sample should have sufficient volume to allow the obtaining of as many samples as are necessary to represent the characteristics of the rock, with at least 15 cm. The fragment to be analyzed, about 5 cm, should be cut in the form of a parallelogram of $4 \times 3 \times 1$ cm ^{3}. One of the faces should be sanded and glued to the glass slide, previously frosted with abrasive.

The parallelogram should then be cut to a thickness of 5 mm, and worn down on the grinding wheel to a thickness of 70 µm, and then ground with progressively finer grain abrasive, until a thin blade of 30 µm thickness is obtained. The surface should then be cleaned with a suitable reagent (xylene or similar) and glued to the coverslip of glass.

The operating principle of a petrographic microscope is the polarization of light. Luminous energy propagates with a transverse wave motion in which the vibration of the particles is perpendicular to the direction in which the energy moves. Thus, light (unpolarized) propagates in directions, forming a right angle with the direction of the luminous radiation. Plane polarized light has its propagation restricted to a single direction in space. The materials or devices that transform conventional light into plane-polarized light are called polarizers (Fig. 6.3).

The optical polarization microscope has a polarizing crystal below the stage that transmits vertically polarized light, and another polarizer, called an analyzer, mounted on the tube located above the stage, which transmits light in the transverse direction (Figs. 6.2 and 6.4).

The characteristics to be observed in the thin section are: (i) texture; (ii) mineralogical composition with indication, in percentage, of essential and accessory minerals; (iii) nature and classification of the rock; (iv) microfissural state; (v) state and type

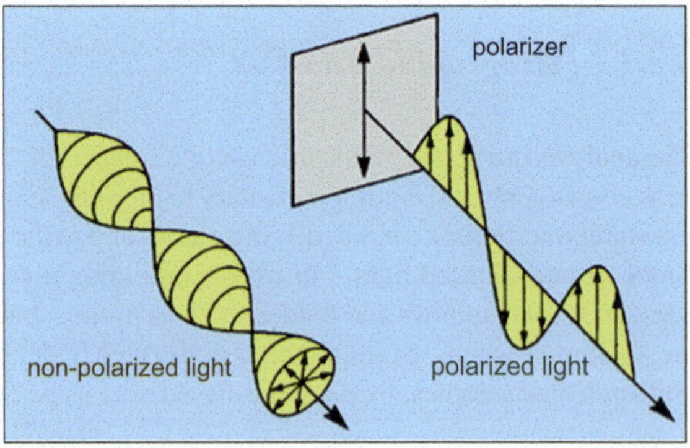

Fig. 6.3 Polarization of light. Conventional light and polarized light

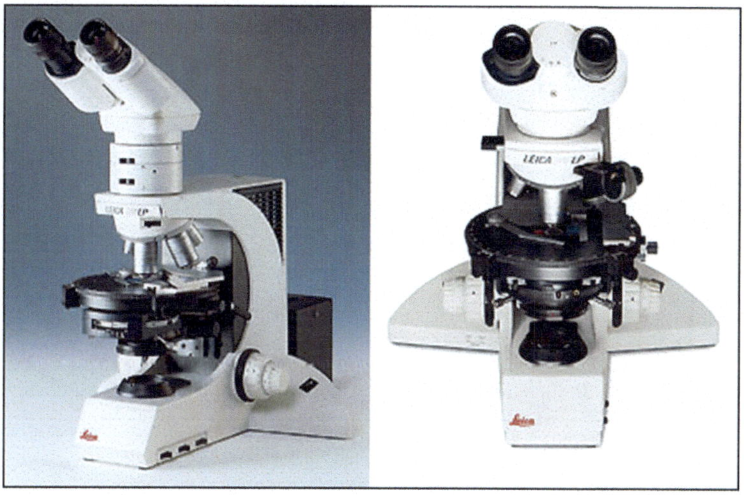

Fig. 6.4 Optical microscope (or petrographic microscope) of polarization for petrographic analysis

of alteration of the minerals, categorizing the degree of alteration of the rock; and (vi) presence or absence of deleterious minerals.

Petrography in thin sections is the best way to identify silicates, such as those of the feldspar, pyroxene, amphibole and mica groups, which have great compositional or structural similarity and, therefore, difficulties for their identification by SEM/EDS or XRD. The quantification of phases, on the other hand, is more restricted; the classic modal analysis, manual point counting with identification of the phase that appears in the center of the field of view with fixed increment of the stage, is a time-consuming analysis, which requires a specialized operator. Figure 6.5 presents a comparison between the images of pyroxene by conventional optical microscopy (natural light) and that obtained by petrographic microscope (polarized light).

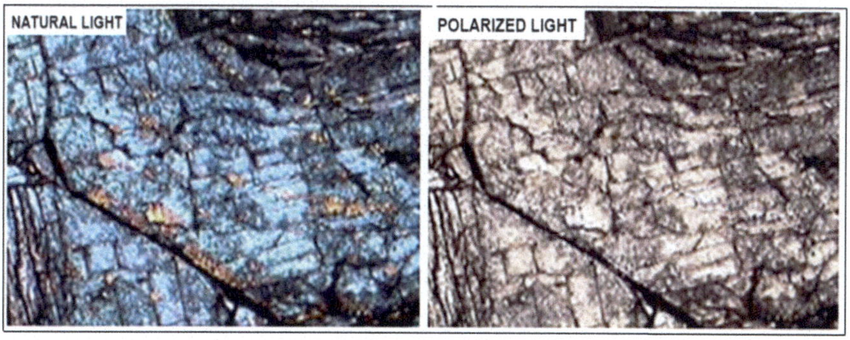

Fig. 6.5 Images of pyroxene by conventional optical microscope (natural light) and petrographic microscope (polarized light)

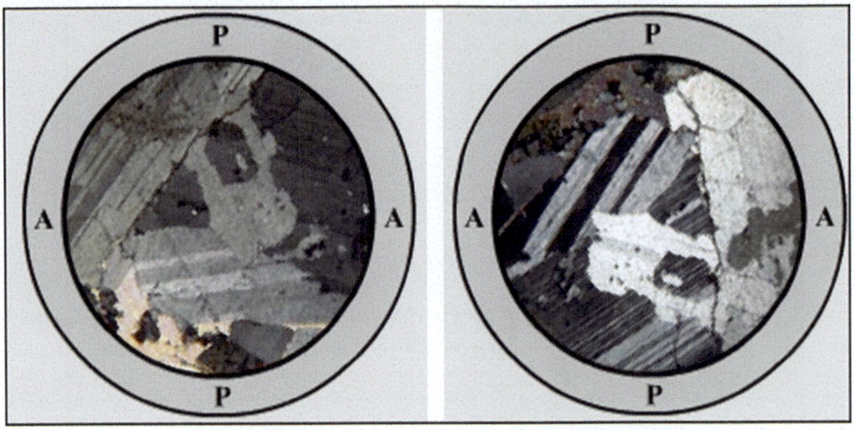

Fig. 6.6 Multiple twinning of plagioclase under a petrographic microscope

Since the optical properties for the identification of most transparent minerals depend on the geometry of the crystal intersection and the angle of incidence of light, the colors vary with the rotatory movement of the stage, so that the automatic recognition of minerals, by image analysis, is impractical at the moment, with rare exceptions.

Figure 6.6 presents images of petrographic analysis of plagioclase with the formation of multiple twins, which appear clearly, for example, with a rotation of 180º around one of the crystallographic axes.

Some minerals exhibit pleochroism, which is the phenomenon of color variation depending on the orientation of the crystal under light incidence. In general, this is weak. It is a very useful recognition criterion for some minerals, such as biotite (Fig. 6.7).

Another example is birefringence or the presence of high or low interference. Figures 6.8 and 6.9 respectively present images under a petrographic microscope of high interference, as is the case with muscovite, or low interference, as is the case with quartz.

6.3 Electron Microscopy

The resolution limit of an optical microscope is determined by the wavelength of the light used to illuminate the sample. The resolution limit can decrease if radiation with a shorter wavelength (λ) is used. The wavelength of a beam of strongly accelerated electrons is many orders of magnitude smaller than visible light and even smaller than ultraviolet light (0.004 nm at 100 kV compared to 800–200nm of visible and UV light). The advantage that electrons have over other small λ radiations is that

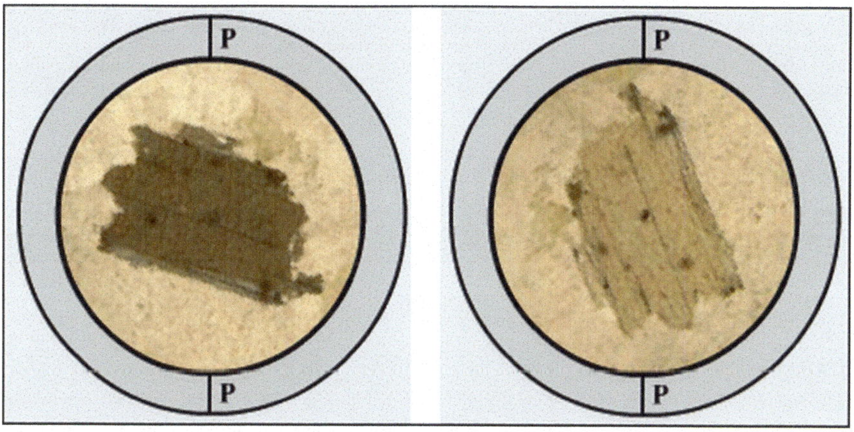

Fig. 6.7 Pleochroism of biotite under a petrographic microscope

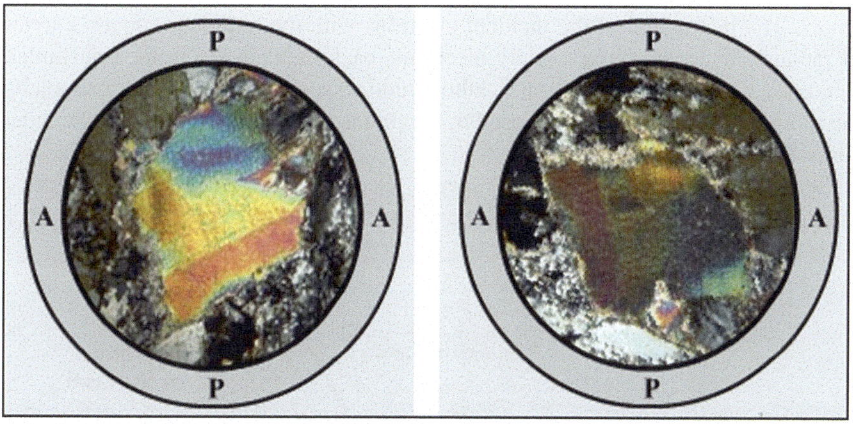

Fig. 6.8 High interference colors in muscovite under a petrographic microscope

electrons are easily accelerated by a potential difference and their trajectory can be modified by the action of electric or magnetic fields.

The need to work in a vacuum (10^{-7} to 10^{-10} bar) is to prevent the electrons in their trajectory towards the sample from being interfered with by atoms or molecules present in the atmosphere.

There are different types of electron microscopes, but the most well-known are the scanning electron microscope (SEM) and the transmission electron microscope (TEM). Both differ greatly in operating principles as well as in the results obtained with their use. In addition, a combination of the two, the STEM (scanning TEM) allows for the analysis of a thin sample by scanning.

The basis of electron microscopy is the use of a beam of accelerated electrons as a radiation source and its focusing on the surface of the sample using condenser

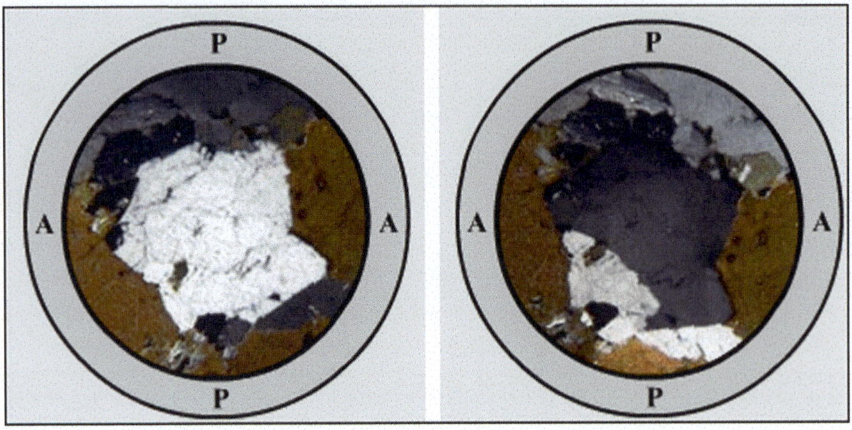

Fig. 6.9 Low interference colors in quartz under a petrographic microscope

lenses. The interaction of the incident electrons with the sample produces a series of radiations secondary: secondary electrons, backscattered electrons, transmitted electrons, X-rays, Auger electrons, cathodoluminescence, and energy absorption by the sample. These radiations are used to obtain information about the sample under analysis (Fig. 6.10).

Secondary electrons: are low energy electrons (less than 50 eV) ejected throughout the braking trajectory of the beam electrons within the sample. Images

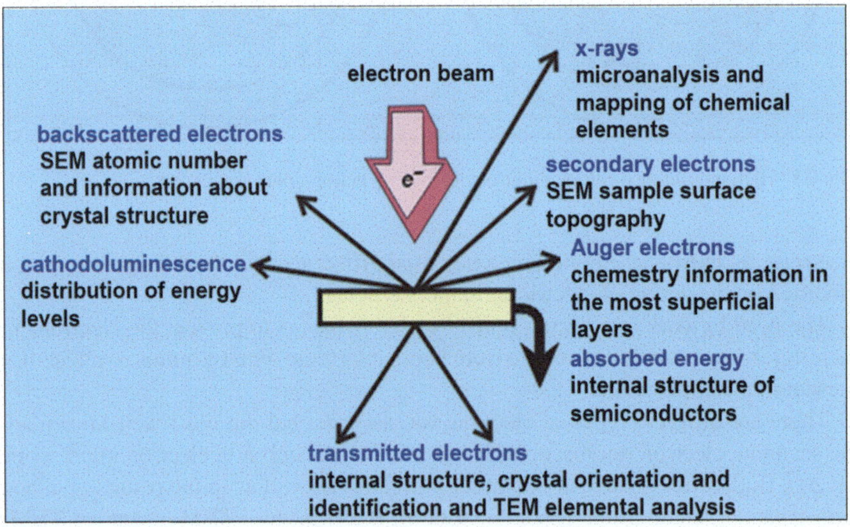

Fig. 6.10 Radiations used in electron microscopy to obtain information about the sample under analysis

obtained with these electrons are the most traditional in SEM and provide information about the topography. They allow high magnifications, but lose in image quality when analyzing poorly conductive samples.

Backscattered electrons: Backscattered electrons originate from the elastic scattering of the beam electrons by the sample atoms. Atoms with a higher atomic number (heavier elements) generate backscattered electrons, so heavier atoms will appear brighter than their neighbors. Images in backscattered electron mode also have a topographic contrast added to the compositional contrast. Images with backscattered electrons do not allow as high magnifications as images with secondary electrons; the magnification limit is around 5000×, but they are efficient in obtaining images of non-conductive samples.

Characteristic X-rays: if the electron beam has high energy (generally 15 keV), it can ionize internal levels of the sample atoms' electron shell. The internal ionization is quickly compensated by relaxation processes accompanied by the emission of X-rays with energy characteristic of each atom. By measuring the energy of the emitted characteristic X-rays, the elements present in the sample are identified using Moseley's Law, which correlates the atomic number of the atoms with the energy of their characteristic X-rays. Maps of this signal allow observing the distribution of the elements in the image.

6.3.1 Scanning Electron Microscopy (SEM)

One of the limitations of optical microscopy is the maximum magnification of around 2000 times. As a result, small structural details cannot be detected by this technique. Scanning electron microscopy (SEM) presents itself as the most suitable technique in these cases, as it allows achieving much higher magnifications than those achieved through the technique of optical microscopy. In SEM, the area or microvolume to be analyzed is irradiated by a thin beam of electrons as opposed to visible light radiation as is the case with optical microscopy.

Figure 6.11 presents the main parts of a scanning electron microscope. The SEM consists of five basic components:

(i) a filament and a high voltage system, which together form the electron gun;
(ii) condensing electromagnetic lenses for beam focusing control;
(iii) objective lens for scanning the beam over sample regions;
(iv) apertures for the collimation of the electron beam;
(v) detectors for the collection of different signals coming from the sample.

The operating principle of the SEM consists of the emission of electron beams by a tungsten capillary filament (negative electrode), through the application of a potential difference that can vary from 0.5 to 40 kV. The voltage variation allows the acceleration of the electrons towards the positive electrode to be controlled. The correction of the beam path is carried out by the condenser lenses that align the beams

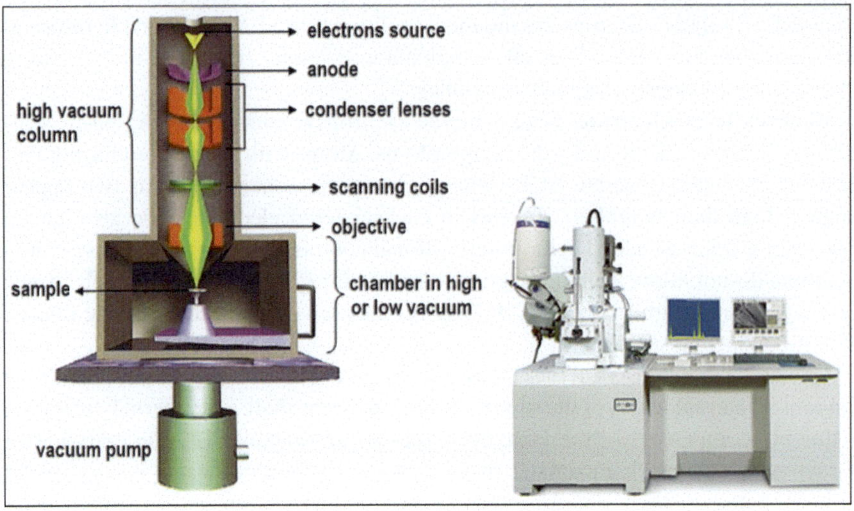

Fig. 6.11 Main parts of a scanning electron microscope and photograph of a commercial equipment

towards the objective aperture. The objective adjusts the focus of the electron beams before the electrons hit the sample under analysis.

The signals of greatest interest for image formation are the secondary electrons and the backscattered ones. As the primary electron beam scans the sample, these signals undergo modifications according to the variations of the surface. The secondary electrons provide an image of the sample surface topography and are responsible for obtaining high-resolution images, while the backscattered ones provide an image characteristic of composition variation.

The SEM has a sample holder capable of performing translation, rotation and tilt movements to position the object and select regions for analysis. In general, it is necessary that this whole set of components is in a vacuum chamber and that the sample also supports vacuum. For samples that dry out in vacuum, it is advisable to use a low vacuum SEM.

The SEM increases its investigative potential with the coupling of an energy dispersive spectrometer (EDS), which allows for point chemical analysis on the sample under observation, identifying chemical elements, except the four lightest (H, He, Li and Be).

6.3.1.1 Energy Dispersive X-ray Spectrometry—EDS

The EDS is an essential accessory in the study of microscopic characterization of materials. When the electron beam hits a mineral, the outermost electrons of the atoms and constituent ions are excited, changing energy levels. When they return to their initial position, they release the acquired energy which is emitted in wavelength in the x-ray spectrum. A detector installed in the SEM vacuum chamber measures the energy associated with this electron. As the electrons of a certain atom have different energies, it is possible, at the point of incidence of the beam, to determine which chemical elements are present at that location and thus identify in moments which mineral is being observed. The reduced diameter of the beam allows for almost point mineral composition determination (<2 μm).

The combined use of EDS with the SEM is of great importance in petrographic characterization and petrological study. While the SEM provides clear images (even though they are virtual, as what is seen on the computer monitor is the transcodification of the energy emitted by the particles, instead of the radiation emitted by light), the EDS allows for its immediate identification. In addition to mineral identification, the equipment also allows the mapping of the distribution of chemical elements by minerals, generating compositional maps of elements with atomic mass greater than that of boron and that are present in concentration greater than 1%.

6.3.1.2 Backscattered Electron Diffraction—EBSD

The energy backscattering diffraction (EBSD) is a technique that consists of placing a sample with a perfectly flat surface inclined at 70° with the incident electron beam. The backscattered electrons generate a diffraction pattern, which appears in the form of streaks (Kikuchi streaks), which can be viewed on a monitor along with the SEM image of the beam incidence site.

The EBSD has been widely used in the microstructural characterization of polycrystalline aggregates of any nature. Its use allows the determination of orientations of any plane or crystallographic direction in very small regions (depending on the width of the SEM electron beam) or throughout the crystalline aggregate. The EBSD, used in conjunction with the *EDS*, allows the identification of any crystalline material from the constituent elements, symmetry, and parameters of the crystal lattice. For the characterization of mineral raw materials, the most important images are those of EBSD. The importance of these images for characterization of ores is the possibility of separating the phases by their gray intensity. Figure 6.12 presents an image of an ore obtained by this technique. Just by the level of gray, 6 different phases are easily identified, and after checking that there are not two phases with the same level of gray (for example, sphalerite and chalcopyrite), even the quantification of each one is possible.

The combination of EBSD and EDS allows to verify the presence of variable composition within the same mineral. Figure 6.13 shows a pyrochlore grain with considerable variation in compositions, presented in the table inserted in the image.

6.3.1.3 Sample Preparation for SEM

Working in high vacuum, the sample surface must be conductive, to attract the electrons from the beam. If the sample is not naturally conductive it must receive a

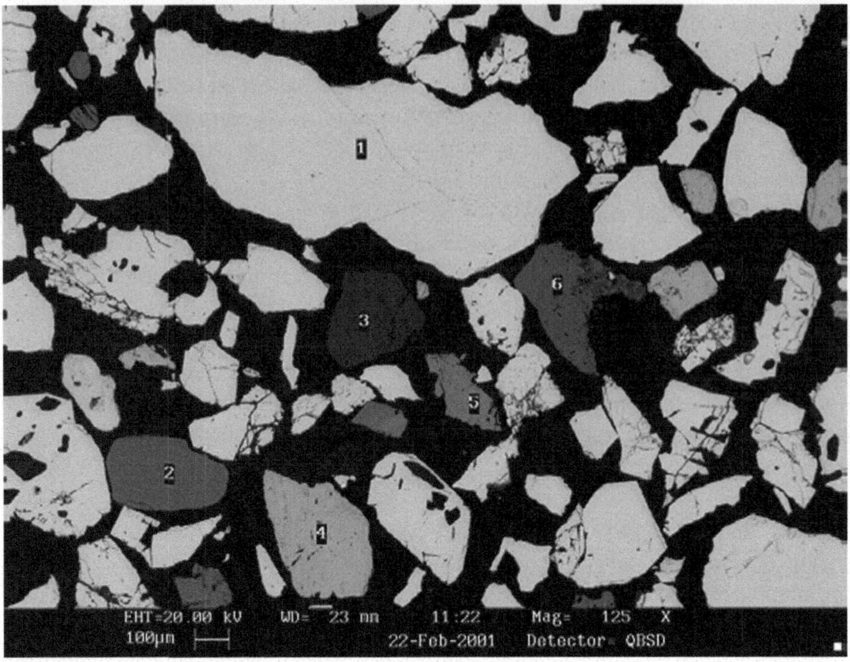

Fig. 6.12 Image obtained with backscattered electron detector in SEM. 1-cassiterite, 2-zircon, 3-gibbsite, 4-barite, 5-mineral from the pyrochlore group (betafite), 6-pyrite (partially altered to iron oxides/hydroxide)

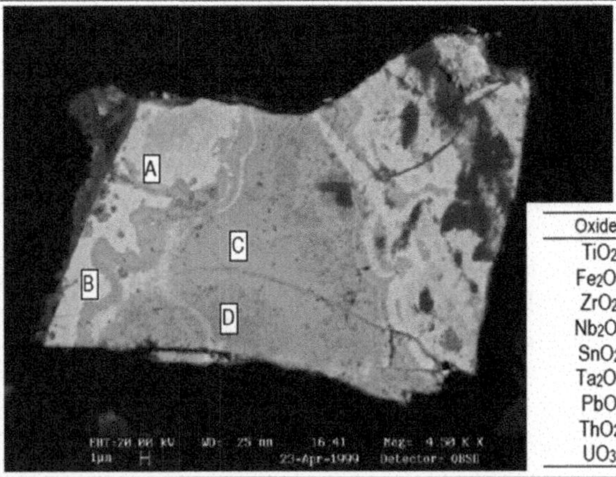

Oxide	A	B	C	D
TiO_2	0,57	0,64	0,45	0,95
Fe_2O_3	1,29	1,04	0,97	1,22
ZrO_2	1,18	0,65	2,28	1,76
Nb_2O_5	35,07	36,22	32,02	32,42
SnO_2	3,53	3,34	1,66	0,47
Ta_2O_5	6,62	6,93	14,36	12,79
PbO	42,97	42,43	35,67	21,33
ThO_2	1,13	0,96	0,66	1,45
UO_3	7,65	7,78	11,92	27,61

Fig. 6.13 Backscattered electron image of uranium-lead-pyrochlore grain, with very varied gray levels due to composition, and semi-quantitative analyses referring to the points indicated in the image

coating, including if the resin is insulating. The most common are carbon or gold by sputtering (physical deposition from vapor phase). In samples with relief, gold is used.

As gold is a better conductor of electrons, it allows images of better resolution and more magnification, but it partially inhibits the response of the backscattered electron detector, and its many peaks in the EDS spectrum interfere with qualitative analyses, and practically make quantitative ones impossible. Carbon, on the other hand, presents a single peak in the EDS spectrum and, therefore, interferes very little in microanalyses. Pure carbon is light, not inhibiting the response of the backscattered electron detector. As it is applied by vaporization in vacuum, however, a good layer can only be achieved on smooth and flat surfaces.

Figure 6.14 presents an example of the difference between the coating with gold or with carbon. The image of the sample coated with gold (Fig. 6.14a), although clearer, does not allow the impurities to be distinguished, in the backscattered electron image, as seen in the image of Fig. 6.14b.

6.3.2 Transmission Electron Microscopy (TEM)

A transmission electron microscope (TEM) consists of an electron beam and a set of electromagnetic lenses, which control the beam, enclosed in an evacuated column with a pressure of about 10^{-5} mm Hg. Figure 6.15 schematically presents the TEM and a commercial equipment.

A modern transmission microscope has five or six magnetic lenses, in addition to several electromagnetic deflection coils and apertures located along the path of the electron beam. Among these components, the following three stand out for their importance with respect to electron diffraction phenomena: objective lens, objective aperture, and selective diffraction aperture. The function of the projector lenses is merely the production of a parallel beam of sufficient intensity incident on the sample surface.

The electrons leave the sample from the lower surface with an intensity and direction distribution mainly controlled by the diffraction laws imposed by the crystalline arrangement of the atoms in the sample. Then, the objective lens comes into action, forming the first image of this angular distribution of the diffracted electron beams. After this very important process of the objective lens, the remaining lenses only serve to enlarge the image or diffraction diagram for future observation on a screen.

Although there are some devices in operation whose acceleration voltage is 1000 kV, most of the equipment used in the study of mineral raw materials has an acceleration voltage of up to 200 kV.

In transmission electron microscopy, the observed image is the projection of a certain thickness of the material, with a difference compared to what is observed on a surface.

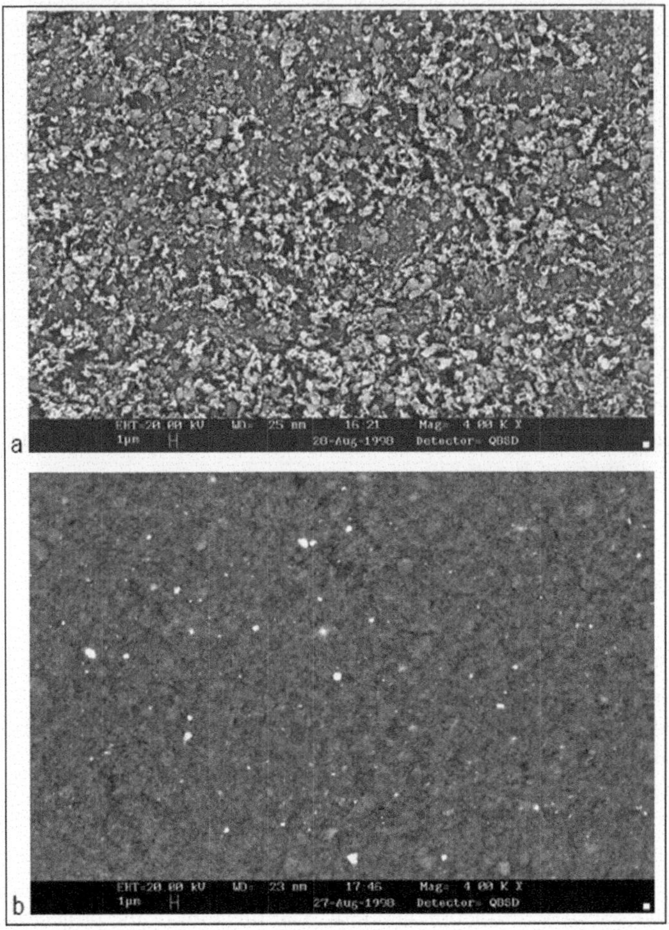

Fig. 6.14 Backscattered electron images of a kaolin fines sample: **a** gold coating, **b** carbon coating

High-resolution microscopes (three image-generating lenses) are capable of magnifying the image up to 500,000 times and have a resolution power of fractions of nm.

There are different modes of image formation in a transmission microscope (Fig. 6.15). The most used is when the image of the object is dark on a bright background (bright field image), as it is formed from the transmitted beam, which has not been scattered. If, on the contrary, scattered electrons are used, the image appears bright on a dark background (dark field image).

On the other hand, with TEM it is possible to obtain a diffraction diagram of the sample, which is valuable information about its crystalline structure. This is possible if the electron beam hits a crystal at an angle capable of satisfying the **Bragg's law** for a certain distance between atomic planes d_{hkl}. The diffraction diagram is formed by

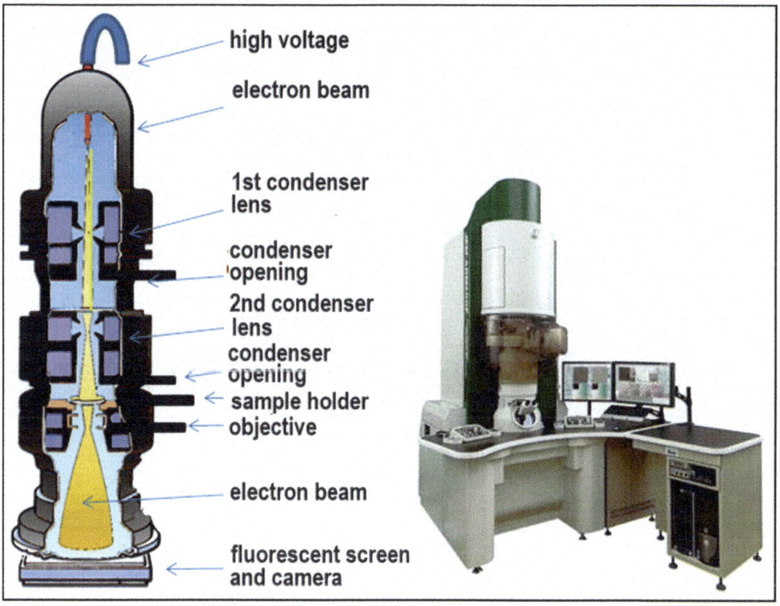

Fig. 6.15 Schematic operation of the TEM and photograph of a commercial equipment

the intersection points of the diffracted beams. The possibility of combining electron diffraction with the different modes of image formation makes the transmission microscope one of the best tools in the study of the crystal lattice and its defects. Figure 6.16 presents the TEM image of high-resolution and the corresponding image with electron diffraction pattern of a Sr_2 $ScFeAsO_3$ crystal along the crystallographic axis [100] (Fig. 6.17).

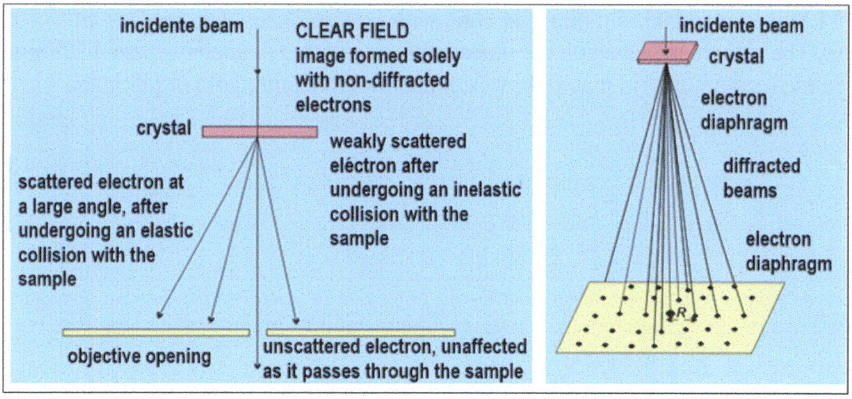

Fig. 6.16 High-resolution image formation in the transmission electron microscope and the corresponding image with electron diffraction patter

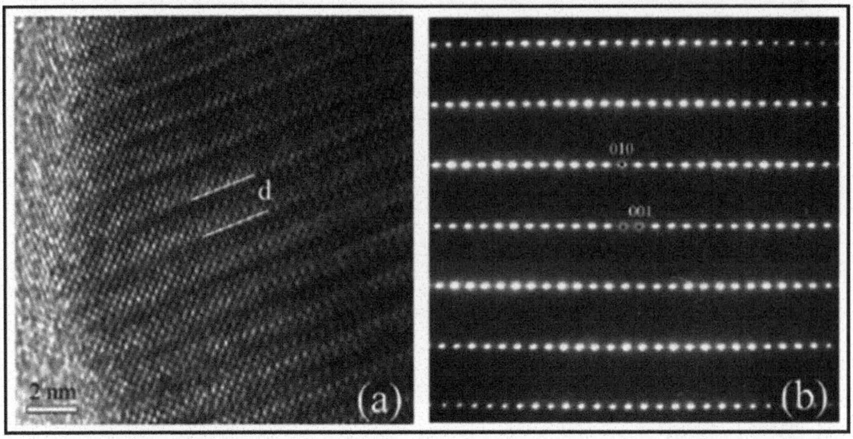

Fig. 6.17 High-resolution TEM image (**a**) and the corresponding image with electron diffraction pattern (**b**) of a $Sr_2\,ScFeAsO_3$ crystal along the crystallographic axis [100]

6.3.2.1 Sample Preparation for TEM

The samples used in TEM should have the following characteristics: thickness from 500 to 5000 Å and, polished and clean surface on both sides. During preparation, the sample should not be altered.

The preparation of thin slices usually follows the following sequence of preparation: cutting slices of 0.8–1.0 mm thick, thinning by mechanical polishing to 0.10–0.20 mm thick. The final thinning of the sample slices of mineral raw materials is usually done by ion milling. Samples transparent to electrons have a thickness of about 100 nm, but this value depends on the acceleration voltage.

The sample holder of a TEM has a standard grid size on which the sample is placed (Fig. 6.18) or a standard size of self-sustaining sample. The standard size of a TEM grid is a 3.05 mm diameter ring, with size, thickness and meshes up to 100 μm. The sample is placed on the inner mesh area with a diameter of about 2.5 mm. The most common grid materials are copper, molybdenum, gold or platinum.

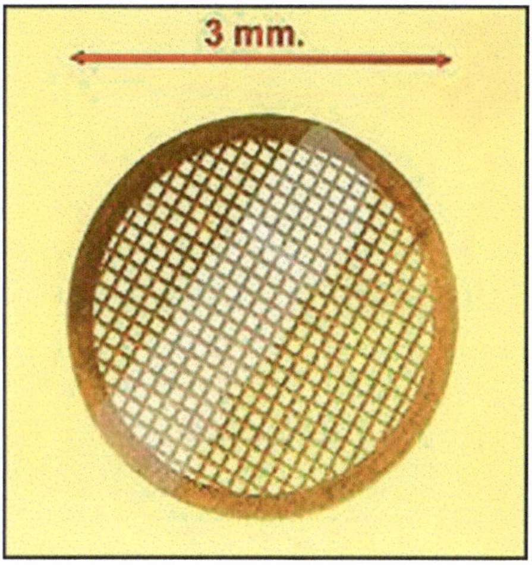

Fig. 6.18 Grid used to support the sample for TEM analysis

Annex I

Tables of IR absorptions in organic molecule bonds. The absorptions are recorded in *wave numbers* (cm^{-1}).

Bond	Type of bond	Specific type of bond	Absorption range and intensity
C–H	Alkyl	Methyl	1380 cm^{-1} (weak), 1460 cm^{-1} (strong) and 2870, 2960 cm^{-1} (both medium to strong)
		Methylene	1470 cm^{-1} (strong) and 2850, 2925 cm^{-1} (both medium to strong)
		Methine	2890 cm^{-1} (weak)
	Vinyl	$C{=}CH_2$	900 cm^{-1} (strong) and 2975, 3080 cm^{-1} (medium)
		C=CH	3020 cm^{-1} (medium)
		Alkenes monosubstituted	900, 990 cm^{-1} (both strong)
		Cis-disubstituted alkenes	670–700 cm^{-1} (strong)
		Trans-disubstituted alkenes	965 cm^{-1} (strong)
		Trisubstituted alkenes	800–840 cm^{-1} (medium to strong)
	Aromatic	Benzene	3070 cm^{-1} (weak)
		Monosubstituted benzene	700–750 cm^{-1} (strong) and 700 ± 10 cm^{-1} (strong)
		Ortho-disubstituted benzene	750 cm^{-1} (strong)
		Meta-disubstituted benzene	750–800 cm^{-1} (strong) and 860–900 cm^{-1} (strong)

(continued)

C. P. Bergmann, *Techniques for Analysis of Mineral Raw Materials*, Topics in Mining, Metallurgy and Materials Engineering, https://doi.org/10.1007/978-3-031-58985-0

(continued)

Bond	Type of bond	Specific type of bond	Absorption range and intensity
		Para-disubstituted benzene	800–860 cm^{-1} (strong)
	Alkynes		3300 cm^{-1} (medium)
	Aldehydes		2720, 2820 cm^{-1} (medium)
C–C	Acyclic C–C	Monosubstituted alkenes	1645 cm^{-1} (medium)
		1,1-disubstituted alkenes	1655 cm^{-1} (medium)
		Cis-1,2-disubstituted alkenes	1660 cm^{-1} (average)
		Trans-1,2-disubstituted alkenes	1675 cm^{-1} (average)
		Tri and tetrasubstituted alkenes	1670 cm^{-1} (weak)
	C–C conjugated	Dienes	1600, 1650 cm^{-1} (strong)
		With benzene rings	1625 cm^{-1} (strong)
		With $C = O$	1600 cm^{-1} (strong)
	C=C aromatic		1450, 1500, 1580, 1600 cm^{-1} (weak to strong)—always all 4!
	C–C triple	Terminal alkynes	2100–2140 cm^{-1} (weak)
		Disubstituted alkynes	2190–2260 cm^{-1} (very weak, sometimes not visible)
C=O	Aldehyde/ketone	Saturated aliphatic/ cyclic of 6 members	1720 cm^{-1}
		α, β-unsaturated	1685 cm^{-1} (also for terminal ketones)
		Cyclic of 5 members	1750 cm^{-1}
		Cyclic of 4 members	1775 cm^{-1}
		Aldehydes	1725 cm^{-1} (influence of conjugation as with ketones)
	Carboxylic acids and their derivatives	Saturated carboxylic acids	1710 cm^{-1}
		Unsaturated carboxylic acids and aromatic carboxylic acids	1680–1690 cm^{-1}
		Esters and lactones	1735 cm^{-1} (influence of conjugation and ring size as in ketones)
		Anhydrides	1760 and 1820 cm^{-1}
		Halides	1800 cm^{-1}

(continued)

(continued)

Bond	Type of bond	Specific type of bond	Absorption range and intensity
		Amides	1650 cm^{-1} (associated amides)
		Carboxylates (salts)	1550–1610 cm^{-1} (also for zwitterions of amino acids)
O–H	Alcohols, phenols		3610–3670 cm^{-1} (concentrating the samples widens the band and moves it to 3200–3400 cm^{-1})
	Carboxylic acids		3500–3560 cm^{-1} (concentrating the samples broadens the band and moves it to 3000 cm^{-1})
N–H	Amines primary		doublet between 3400 and 3500 cm^{-1} and 1560–1640 cm^{-1} (strong)
	Secondary amines		above 3000 cm^{-1} (weak to medium)
	Ions ammonium		broad band with multiple peaks between 2400 and 3200 cm^{-1}
C–O	Alcohols	Primary	1050 ± 10 cm^{-1}
		Secondary	Around 1100 cm^{-1}
		Tertiary	1150–1200 cm^{-1}
	Phenols		1200 cm^{-1}
	Ethers	Aliphatic	1120 cm^{-1}
		Aromatic	1220–1260 cm^{-1}
	Carboxylic acids		1250–1300 cm^{-1}
	Esters		1100–1300 cm^{-1} (two bands—distinction from ketones, which do not have C–O!)
C–N	Aliphatic amines		1020–1220 cm^{-1} (often overlapped)
	C=N		1615–1700 cm^{-1} (similar conjugation effects to C=O)
	Nitrile (C≡N bond)		2210–2260 cm^{-1} (non-conjugated 2250 cm^{-1}, conjugated 2230 cm^{-1})
	Isonitriles (R-N–C bond)		2165–2110 cm^{-1} (2140–1990 cm^{-1} for R–N=C=S)

(continued)

(continued)

Bond	Type of bond	Specific type of bond	Absorption range and intensity
C–X (X = F, Cl, Br, I)	Fluoroalkanes	Ordinary	1000–1100 cm^{-1}
		Trifluoromethyl	two strong, broad bands between 1100 and 1200 cm^{-1}
	Chloroalkanes		540–760 cm^{-1} (medium to weak)
	Bromoalkanes		Below 600 cm^{-1}
	Iodoalkanes		Below 600 cm^{-1}
N–O	Nitro compounds	Aliphatic	1550 cm^{-1} (strongest band) and 1380 cm^{-1} (weaker band)—always both!
		Aromatics	1520, 1350 cm^{-1} (conjugation usually reduces the wave number)

Glossary

Absorption Relationship between the mass absorbed from the saturating liquid by a test body until saturation (wet mass) and the mass of the dry test body (dry mass).

Adsorption Process in which the molecules of a fluid are concentrated on the surface of a material by chemical, physical or both forces.

Alumina Aluminum oxide, with the formula Al_2O. It exists in various crystalline forms, the main ones being alpha-alumina and gamma-alumina. It is used as a raw material in the refractory industry, mainly in the manufacture of aluminous refractories.

Asbestos (asbestos) Fibrous mineral of varied chemical composition, mainly composed of hydrated silicates of magnesium, such as: chrysotile, actinolite, tremolite, antigonite and serpentine.

Anatase A mineral with the formula TiO_2, and different crystallization from rutile.

Andalusite A mineral with the empirical formula $Al_2O_3.SiO_2$, with a crystalline structure different from the minerals kyanite and sillimanite, which have the same formula.

Anode The negative electrode of an electrochemical circuit (see electrode).

Apatite A hexagonal mineral with the composition $3Ca_3\ (PO_4)_2$ with a hardness of 5 on the Mohs scale.

Asbestos See asbestos.

Analysis Decomposition of a whole into its parts constituents, in order to identify each of these parts.

Apparent mass density Relationship of the mass of the test body when dry to the apparent volume of the same, commonly expressed in grams per cubic centimeters.

Apparent mass density of the solid part Relationship of the mass of the test body when dry to the apparent volume of the solid part of the same, commonly expressed in grams per cubic centimeters.

Attribute inspection Method that consists in noting, in each of the items of a lot or of a sample drawn from this lot, the presence or absence of a certain qualitative characteristic (attribute) and in counting how many items have or do not have

C. P. Bergmann, *Techniques for Analysis of Mineral Raw Materials*, Topics in Mining, Metallurgy and Materials Engineering, https://doi.org/10.1007/978-3-031-58985-0

this characteristic. In summary, attribute inspection simply classifies the item as defective or non-defective, according to a certain criterion. See attribute.

Apparent mass See apparent mass density.

Apparent mass of the solid part See density of apparent mass of the solid part.

Apparent porosity Ratio of the volume of open pores to the apparent volume of the test body, expressed in percentage.

Apparent volume Volume of a piece or a test body, including the volume of the solid part and that of the open and closed pores.

Apparent volume of the solid part Volume of a piece or a test body, including the volume of the solid part and that of the closed pores.

Ash Non-combustible solid residue that remains after the burning of any combustible material.

Amorphous Absence of crystalline structure or of defined molecular arrangement, without defined form.

Agalmatolite Metamorphic rock essentially made up of pyrophyllite.

Attribute A quality characteristic whose inspection is carried out only by noting its presence or absence. In general, the attribute is a qualitative property, not measurable, or whose measurement is not practical; examples include appearance defects (visual aspect).

Baddeleyite Mineral composed essentially of zirconium dioxide.

Batch Quantity of material manufactured in a production sequence for which uniformity in specific characteristics is expected.

Bauxite Residual rock composed essentially of aluminum hydroxides, such as: diaspore, gibbsite, boehmite, and impurities like iron hydroxides, quartz and clay minerals.

Bentonite Rock composed essentially of montmorillonite clay, with high plasticity.

Beryllia Beryllium oxide, with the formula BeO.

Beryl Double silicate of beryllium and aluminum, with the formula $3BeO.Al_2O_3.6SiO_2$, from which beryllium oxide is obtained.

BET Instrumental method for determining the surface area of a solid sample, measuring the monomolecular adsorption of nitrogen at the temperature of liquid nitrogen (210 °C below zero).

Boehmite or boehmita Aluminum hydroxide with the empirical formula AlO(OH) or $Al_2O_3.H_2O$, usually found in bauxites.

Borax Sodium tetraborate with the empirical formula $Na_2O_3.2B_2O_3.10H_2O$, used as an additive in the manufacture of refractories.

Brucite Mineral with the formula $Mg(OH)_2$.

Boric acid Chemical product generally used as a sintering agent in ramming refractory masses, with the formula H_3BO_3.

Congruent fusion Change of a substance, when heated, from the solid state to the liquid state of the same composition. Example: Melting of ice.

Caustic magnesia See calcined magnesia.

Calcined magnesia Magnesia obtained by calcination of magnesite by heating to a temperature not exceeding 1200 °C.

Calcined refractory bauxite Low iron, titanium and other impurities bauxite, calcined, used for manufacturing refractories.

Crucible Bottom part of furnaces, where molten metals and slags accumulate. When referred to a refractory piece, it is a piece of truncated cone shape, with the smaller base closed and used to melt and/or contain materials in processing.

Combined water Water chemically held as water of crystallization.

Calcined or bayer alumina Alumina obtained by the bayer process.

Chemical analysis Set of procedures aimed at determining the composition of a refractory material usually expressed in terms of oxides.

Combustion air Air used in a burner for burning fuels.

Crystal lattice Regular, repeated and geometric arrangement of atoms, ions, or molecules in a crystal.

Calcined magnesite See calcined magnesia.

Clay A generally plastic rock, made up essentially of hydrated aluminum silicates, called clay minerals, which may contain free silica and other impurities within its crystalline network, such as magnesium and iron.

Conductibility or conductivity Property that bodies have to be conductors of heat, electricity, sound, etc.

Cryolite Material of formula Al_2F_6, synthetically produced and used as a catalyst in the reduction of alumina in the production of primary aluminum.

Cryptocrystalline Crystalline structure in which the individual crystals are small and cannot be seen through petrographic microscopes, but through an electron microscope; many of the so-called amorphous minerals are, in reality, cryptocrystalline.

Crystal Homogeneous solid body of an element compound or isomorphic mixture, having a regularly repeated atomic arrangement and which can be externally expressed by flat faces.

Crystallization Solidification of certain substances minerals that take defined geometric forms.

Cristobalite Crystalline polymorphic variety of silica stable above 1450 °C.

Chromite Chromium spinel of empirical formula $FeO.Cr_2O_3$; essential component of chromium ores. The term chromite is used to designate chromium ore. The other constituents of chromite are a mixture of spinels of composition $(FeO.MgO).Al_2O_3.Fe_2O_3$ and siliceous gangue.

Clay mineral Minerals that make up clays.

Coefficient of thermal conductivity (or conductivity) Amount of heat per unit of time that flows through the unit surface of a test body, per unit of thickness, when subjected to a temperature variation of one degree. It is commonly expressed in kcal.m/m^2.h.°C. In the international system of units (SI), it is expressed in W/(m.K). To convert the unit from kcal.m/m^2.h.°C to W/(m.K), multiply by 1.162279.

Colloid Any fine material that can be easily suspended (dry or wet).

Combustion Process of combining a substance with oxygen (oxidation), usually with an exothermic and self-sustaining reaction; also known as the burning process.

Complete combustion Combustion that combines the appropriate chemical proportions of fuel and air, in such a way that fuel and oxygen are completely consumed.

Comminution Process of reducing the size of materials by division into smaller sizes. The main processes of comminution are crushing and grinding.

Compaction Act, operation or effect of reducing the volume of voids, in order to increase their apparent density.

Components Constituents that are part of the compositions.

Chalcedony Variety of fibrous or concretionary silica.

Calcination Thermal treatment to which some raw materials for the manufacture of refractories are subjected, as a preliminary preparation for processing or subsequent use, with the purpose of eliminating free or chemically combined volatile constituents and to produce volume changes aimed at product stability.

Calcite Mineral with the chemical formula $CaCO_3$ and Mohs hardness 3.

Caldasite Rock essentially composed of badeleyite and zirconite in varying proportions.

Caliophyllite or kaliophyllite Mineral with the formula $K_2O.Al_2O_3.2SiO_2$.

Calorie Amount of heat necessary to increase the temperature of 1g of water in the range of 3.5–4.5 °C (1 cal = 4.1868 j).

Catalyst Substance that causes or accelerates a chemical exchange, without participating ponderously in the reaction.

Chamotte or grog Mixture of refractory clays and other refractory materials, agglomerated and calcined at high temperatures to be used in the manufacture of refractory products.

Ceramics A generic term applied to all materials made from clay and/or generally non-metallic minerals, by the application of heat. Art or technique of producing by a ceramic process. By use, the term has been extended to other materials or activities and currently also includes: refractories, cement, lime, glass, plaster, frits, glazes, abrasives, electrical insulators and thermal insulators produced from non-metallic inorganic materials.

Closed pore Pore not permeable to fluids.

Cathode Positive electrode of an electrochemical circuit.

Combustion chamber Confined space in which the fuel is burned.

Capillarity Ability of refractories to conduct liquids through their open pores by the force of surface tension.

Differential thermal analysis—DTA An instrumental method for graphing the temperature of a sample while its container is heated at a constant rate. It reveals endothermic and exothermic processes occurring in the sample, for example: decomposition, phase changes, combustion.

Dolomitic limestone Limestone containing a certain proportion of dolomite.

Densification See compaction.

Decarburization Process of loss or reduction of carbon.

Dephosphorization Metallurgical process aimed at reducing the phosphorus content of liquid metals.

Devitrification Change from the vitreous condition to crystalline.

Devitrify To convert, partially or completely, from a vitreous state to a crystalline state, usually by controlled heating.

Dead-burned magnesite See sintered magnesia.

Dead-burned magnesia See sintered magnesia.

Diaspore Crystalline aluminum hydroxide of empirical formula $Al_2O_3.H_2O$, which, after calcined, loses its water of hydration, transforming into alumina, being present in bauxites.

Diatomite Siliceous rock of sedimentary origin with different degrees of consolidation, formed mainly by siliceous skeletons of the plant microorganisms called diatoms (the same as "*Kieselgur*").

Disthene See kyanite.

Doloma Product resulting from the calcination of dolomite.

Dolomite Double carbonate of calcium and magnesium of empirical formula $CaCO_3.MgCO_3$.

Dunite Rock essentially composed of olivine, of empirical formula $(Mg.Fe)_2.SiO_4$.

Elasticity Properties that certain materials have to deform and return to their original size and shape after the removal of stress.

Electrode One of a pair of conductors introduced into an electrochemical cell between which ions in the intervening medium flow in opposite directions. When direct current is used, one electrode or pole is positively charged (anode) and the other negatively (cathode). The electrodes may be separated by a small space in which an arc forms or a spark jumps.

Elutriation Process by which a mixture of particles of different sizes is separated into more or less homogeneous fractions, by sedimentation in the flow of a fluid.

Electrofused mullite Mullite produced by the fusion in an electric furnace of a stoichiometric mixture of calcined alumina and pure quartz.

Eutectic or eutectic equilibrium Composition within a system of two or more crystalline phases, whose melting point is lower than any other combination of the same components.

Extrusion Shaping process in which, by applying pressure, a material in the plastic state is forced through a nozzle, taking the shape of this. This process is used for manufacturing gloves, channels, tubes, ceramic coating, etc.

Electrofused magnesia Magnesia melted in an electric furnace.

Fines The name given to the smallest fractions of a granulometric classification, usually below 1.0 mm.

Flocculation Agglomeration of suspended particles, causing their sedimentation.

Flocculant Electrolyte added to a colloidal suspension to cause the aggregation and sedimentation of particles as a result of the reduction of repulsion between them.

Fluorite Cubic mineral of composition CaF_2 and hardness 4 on the Mohs scale.

Furnace Equipment in which, through heat, physical and chemical transformations are produced in a certain material.

Forsterite Magnesium orthosilicate with empirical formula $2MgO.SiO_2$.

Fused magnesia See electrofused magnesia.

Friable Easily reducible to grains or powder.

Flux Substance that lowers the melting point of a material.

Fusion Thermal process by which a solid charge is completely transformed into a liquid phase.

Fused alumina Alumina obtained by electrofusion.

Fixed carbon Carbon remaining from the thermal treatment of refractories, which have in their composition components and/or binders containing carbon.

Fayalite Mineral with empirical chemical composition $2Fe_2O_3.SiO_2$.

Feldspar Constitutes a series of silicates of aluminum and alkaline or alkaline-earth base, anhydrous and crystallized, forming essential and abundant elements in magmatic rocks. They are named orthoclase, microcline, albite, anorthite, etc. There are potassium feldspars, sodium feldspars, calc-sodium and calcium feldspars.

Granulometric analysis Determination of the distribution of the sizes of non-aggregated particles, obtained by sieving, using a series of standardized test sieves.

Gel A colloidal state composed of interdispersed solids, in which the solid particles or ligaments are interlaced in three dimensions.

Gibbsite Aluminum hydroxide with empirical formula $Al_2O_3.3H_2O$.

Gypsum Dihydrated calcium sulfate, with formula $CaSO_4.2H_2O$, with hardness 2 on the Mohs scale.

Grain Size Linear dimension control of individual grains.

Grinding Comminution process aimed at reducing grains to a desired degree of fineness.

Granulometric fraction Set of grains that, in the sieving process through a series of sieves, passes through the mesh of a certain sieve and is retained in the mesh of the immediately lower sieve. Example: Granulometric fraction 4.76 mm/2.00 mm.

Graphite Hexagonal polymorphic variety of carbon.

Granulometry Term that indicates the relative proportion of particles of different sizes. Usually determined by separation into series of fractions by sieving.

Grain Part of the microstructure of a material with greater physicochemical stability, constituted by the coarsest particles, which may contain one or more crystals of the same or different nature. When referring to the granulometric classification process of a comminuted raw material, the grains constitute granulometric fractions individualized by size ranges. The grains usually have an irregular shape, and their dimensions can vary widely.

Glass Inorganic product melted and cooled to a rigid situation, without crystallization.

Granulometric composition Granulometric composition of the proportion of grains of different sizes in the manufacture of refractories.

Heat Form of energy whose effect is produced by the accelerated vibration of molecules. The intensity of heat in a body is its temperature.

Hydrated lime Material obtained by treating lime with enough water to satisfy its affinity for water under its hydration conditions. It is usually composed of calcium hydroxide or a mixture with magnesium hydroxide.

Hardness of a mineral A mineral is said to be harder than another when one scratches the other and is not scratched by it. The hardness is determined by comparison with standard minerals on the Mohs scale (1—talc; 2—gypsum; 3-calcite; 4—fluorite; 5—apatite; 6—feldspar; 7—quartz; 8—topaz; 9—corundum and 10—diamond).

Halloysite Hydrated aluminum silicate similar in composition to kaolinite, which may contain more water of composition, with empirical formula $Al_2Si_2O_5(OH)_4.2H_2O$ and $Al_2Si_2O_5\ (OH)_4$.

Hematite Iron mineral, with formula Fe_2O_3, crystallized in the alpha polymorphic variety.

Hercynite Mineral from the spinel group, with empirical formula $FeO.Al_2O_3$.

Hydrargillite See gibbsite.

Hydration Chemical combination of a mineral with water.

Hygroscopic material Material capable of absorbing water molecules from the atmosphere.

Incongruent fusion Dissociation of a compound on heating with the formation of another compound and a liquid of different composition from the original compound.

Ilmenite Iron titanate with empirical formula $FeO.TiO_2$.

Inspection Conformity assessment activity by measurement, observation, examination, testing, calibration, of one or more appropriate characteristics, and the comparison of the results with specified requirements, in order to determine whether conformity for each of these characteristics is achieved.

Inversion Physical structure change between two or more forms of polymorphic crystals. Example: Inversion from alpha quartz to beta quartz at a temperature of 573 °C.

Isomorphous mixture Solid solution in which mineral compounds of similar chemical composition and crystallization crystallize together in varying proportions.

Isomorphs Minerals of analogous chemical analysis and similar crystalline structure. These minerals usually form solid solutions. Example: $2MgO.SiO_2$ (forsterite) and $2Fe.SiO_2$ (fayalite), forming olivine.

Industrial sieve Equipment used in the granulometric classification of crushed and/or ground raw materials, consisting of one or more screens, usually metallic or of polyurethane.

Instrumental chemical analysis Set of procedures that, in addition to purely chemical methods, involves methods instrumentals, such as: X-ray fluorescence, atomic absorption spectrophotometry, flame photometry, and colorimetry.

Incomplete combustion Partial burning of the fuel–air mixture usually as a result of an inadequate supply of oxygen or at such a low temperature that the chemical reaction does not complete and carbon monoxide is formed instead of carbon dioxide.

Inspection plan Document that sequentially lists inspection activities, including stopping points, organizations and personnel involved, procedures, standards and other documents to be used in the inspection.

Inversion point Temperature at which a reversible crystalline transformation occurs.

Kaolin Clay essentially composed of the clay mineral kaolinite, with the empirical formula $AI_2O_3.SiO_2.2H_2O$, generally resulting from the decomposition of feldspar.

Kaolinite Mineral with the formula $AI_2Si_2O_5(OH)_4$. It is the main constituent of kaolin and many clays.

Kyanite Mineral with the empirical formula $Al_2O_3.SiO_2$ and a crystalline structure different from the minerals andalusite and sillimanite of the same formula.

Kieserite Mineral with the formula $MgSO_4.H_2O$, usually obtained from the residual solutions of potassium salt processing, used in the production of magnesia.

Latent heat of fusion Heat supplied by a liquid during freezing to the solid or absorbed by a solid during melting to liquid, without temperature change.

Loose mass density Relationship between the mass of a bulk material and the volume occupied by this material at rest within a container. It is generally used to determine storage capacity and density of fine powders (below 1 mm).

Lime Calcium oxide resulting from the calcination of limestones and/or other materials composed essentially of calcium carbonate.

Limestone Sedimentary rock composed essentially of calcite.

Latent heat of vaporization Heat supplied by a condensed vapor to liquid or absorbed for liquid evaporating to vapor, without temperature change.

Liquid Temperature System temperature at which there is 100% liquid phase.

Laboratory or test sieve Equipment used in granulometry tests, usually consisting of a square mesh screen fixed to a frame.

Loss on ignition Weight loss resulting from heating a sample of matter at high temperature, after it has undergone a drying process.

Lower heating value (LHV) Amount of heat released by the combustion of 1 kg or $1m^3$ of fuel, referred to normal temperature and pressure conditions, expressed in Joules per kilogram or Joules per cubic meter.

Melting point Temperature at which melting occurs. Generally, refractories do not have a defined melting point—they soften gradually within a temperature range.

Mass density Relationship of a mass to a volume, expressed in kilograms per cubic meters or, commonly, grams per cubic centimeters.

Magnesia Magnesium oxide with empirical formula MgO, natural or synthetic.

Magnesium chromite Spinel with the formula $MgO.Cr_2O_3$.

Magnesium ferrite Mineral from the spinel group with the formula $MgO.Fe_2O_3$.

Magnesium wustite Solid solution of MgO and FeO.

Moisture Content Percentage of water (uncombined) contained in raw materials.

Magnesite Crystalline magnesium carbonate with the empirical formula $MgCO_3$.

Melt Transform a solid into a liquid by applying heat.

Melted alumina Alumina obtained by the fusion of materials containing high levels of aluminum oxide, essentially made up by the mineral corundum (aluminum oxide).

Magnetite Magnetic iron ore, with formula Fe_3O_4.

Mesh Number of openings in a sieve per linear inch.

Mica Group of minerals having perfect basal cleavage and consisting of thin and elastic sheets. The most common varieties are muscovite and biotite.

Modulus of elasticity Measure of the stiffness of a material.

Monticellite Mineral with empirical formula $CaMgSiO_4$, related to olivine.

Montmorillonite Group of clay minerals containing variable proportions of one or more cations of magnesium, potassium, sodium and calcium. It is an essential constituent of bentonites.

Muscovite Mineral from the mica group, with empirical composition $KAI_2(AISi_3)O_{10}(OH)_2$, usually colorless, whitish or light brown, being a common constituent of metamorphic and igneous rocks.

Mullite Crystalline aluminum silicate, with empirical formula $3Al_2O_3.2SiO_2$.

Microsilica See volatilized silica.

Mineral Natural inorganic substance, amorphous or crystalline, which may contain impurities in small proportions and is defined by its chemical composition and physical characteristics.

Nucleation The first phase of the crystallization process.

Nepheline Hexagonal mineral from the feldspathoid group, with composition $(Na.K)Al_5O_4$, product from the scorching of silico-aluminous and aluminous bricks with vapors or slags, containing a high content of soda.

Ore Rock from which a mineral can be economically extracted.

Olivine Mineral with empirical formula $(MgFe)_2SiO_4$, consisting of isomorphic solid solutions of forsterite-fayalite.

Opal Material consisting of silica gel, with empirical formula $SiO_2.nH_2O$.

Oxidation Chemical reaction of an element with oxygen or with compounds containing oxygen.

Oxidation–reduction Chemical reaction in which one of the reactants is reduced (gains one or more electrons) and the other is oxidized (loses one or more electrons).

Oxidizing agent Agent that produces oxidation.

Oxide Binary compound of oxygen with some other element or with a radical.

Oxidizing atmosphere The atmosphere of a furnace with an excess of oxygen and which tends to oxidize materials placed in it.

Oxidizing flame Flame resulting from the combustion of a mixture containing excess air or little fuel.

Open pore Pore permeable to fluids, i.e., that can be filled by fluids.

Phosphoric acid Chemical product generally used as a chemical setting binding agent, with the formula H_3PO_4.

Phase diagram Graphical representation of the equilibrium conditions between the various phases of a substance, through which one can know, under the most varied conditions of temperature, the quantities of the liquid and solid phases of that same substance. Depending on the number of substances that make up the diagram, it is said to constitute binary, tertiary systems, etc.

Poise Unit of measure of viscosity, equivalent to 1g/cm.s (not considered as an SI unit).

Polycrystalline Material formed of several closely linked crystals.

Polymer Compound formed by successive agglomerations of a large number of fundamental molecules.

Polymerization Process in which two or more molecules of the same substance or two or more identical atomic groups join to form a structure of generally high molecular weight.

Polymorphism Property by which certain materials of the same chemical composition form two or more minerals of different crystalline structure. Example: $Al_2O_3.SiO_2$ (sillimanite, andalusite, kyanite).

Pellet Spherical agglomerate, obtained in rotating discs with an axis inclined to the vertical.

Pelletization One of the agglomeration processes.

Periclase Magnesium oxide (MgO) crystallized in the cubic system.

Permeability The ability of the refractory to allow the passage of liquid fluids and gases, through its open pores, expressed in centi-Darcy (cD).

pH Collogarithm (negative logarithm) of the concentration of the hydrogen ion. The pH scale, with values from 0 to 14, is used to measure the acidity or alkalinity of a solution. The value 7 indicates neutrality; values above indicate alkalinity and values below indicate acidity.

Pyrophyllite Hydrated aluminum silicate, crystalline, with empirical formula $Al_2Si_4O_{10}(OH)_2$.

Phase A physically homogeneous portion, but can be mechanically separated from a system.

Pig Iron The name given to an iron-carbon alloy, with a carbon content greater than 1.7% and containing impurities, mainly silicon, sulfur and phosphorus; it is obtained from the reduction of iron ores in electric reduction blast furnaces.

Pre-reacted grains Granular materials resulting from the sintering of an intimate mixture of two or more constituents, finely ground and agglomerated.

Pore Void between the solid particles of the refractory.

Pressing Shaping process through presses.

Pressure Force per unit area, expressed in pascal (Pa), where $1\ Pa = 1N/m^2$ or $1\ MPa = 1\ N/mm^2$.

Procedure Specified way of performing a task. In many cases, these procedures are documented.

Pulverization Reduction of a granulated material to fine particles or powder.

Qualitative analysis Analysis in which some or all of the components and or properties of a sample are identified.

Quartering Equipment used to quarter samples of materials and/or raw materials.

Quartzite Metamorphic rock composed essentially of quartz.

Quartz One of the polymorphic crystalline varieties of silica stable at temperatures below 870 °C, with hardness 7 on the Mohs scale.

Quality plan Document that establishes practices, resources and the sequence of activities related to the quality of a product, a project or a specific contract.

Residual carbon See fixed carbon.

Retained carbon See fixed carbon.

Reducing atmosphere The atmosphere of a furnace with an insufficient amount of oxygen.

Refractory or kaolinitic clay Clay made up essentially of the clay mineral kaolinite.

Reducing flame Flame resulting from the combustion of a mixture containing excess fuel or little air.

Real density Relationship of the mass of the ground substance the powder by the true volume of the solid, commonly expressed in grams per cubic centimeters.

Reversible linear thermal expansion Property of a material to transduce into an increase in its dimensions (length) as a function of temperature increase: the phenomenon is reversible and the result is expressed in percentage.

Reversible thermal expansion Increase in dimensions and volume of materials when heated, which is compensated by a contraction of the same magnitude when they are cooled to the initial temperature. This property is expressed in percentage with reference to a defined temperature.

Reaction Chemical exchange, that is, an exchange where one or more chemical substances (reactants) are consumed and one or more chemical substances appear.

Reactivity Susceptibility of a material to action or chemical change, usually due to a state of subdivision, surface area, metastable form, composition, etc.

Recrystallization Usually a physical process in which, under the influence of external agents, there occurs: growth of one crystal at the expense of another or smaller crystals of the same substance; change of crystalline phase (or crystal lattice or crystal system).

Reduction Chemical exchange in which an atom or ion gains electrons.

Rheology Science that investigates the properties and the mechanical behavior (deformation and flow) of materials.

Repeatability Standard deviation of the results obtained by the same operator, using the same instrument in successive measurements on the same sample.

Reproducibility Standard deviation of the results obtained by different operators, using the same or different instruments in different laboratories on the same sample, using the same method or analytical technique.

Residue Amount of material that remains retained in a mesh after a granulometric analysis.

Resistance to hydration Property defined by the ability of a material to resist chemical combination with water.

Resistance to oxidation Property defined by the ability to resist oxidation.

Reversible Transformation Transformation of the crystalline structure of a material with immediate reversal of direction, caused by a variation in temperature. Example: The transformation of α quartz to β quartz at 573 °C.

Relative Humidity Ratio of the weight of water vapor per unit volume of air to that which the air would have at the same temperature if it were saturated.

Real or true volume Volume only of the solid part of a material, excluding the volume of open and/or closed pores.

Sample Set of one or more items taken from a lot, intended to provide information about it, usually serving as a basis for a decision about the lot.

Sampling Procedure used to extract or constitute a sample.

Soda ash Name given to sodium carbonate, with the formula Na_2CO_3.

Specific heat Amount of heat expressed in joules, necessary to raise the temperature of 1 K for 1 kg of a given substance.

Silicon carbide Synthetic substance of empirical form SiC.

Specification Document that defines requirements and should refer to or include drawings, models or other appropriate documents, as well as indicate the means and criteria by which compliance should be reverified.

Spinel Common designation for minerals of the spinel group, essentially consisting of aluminates of magnesium, with magnesium being replaceable, in varying proportions, by iron manganese or zinc, and aluminum, partially, by iron or chrome. The spinels normally used in the manufacture of refractories are those of chrome (chromite) and those of aluminum.

Steatite (soapstone) Metamorphic rock essentially made up of talc.

Structure Term normally used in conjunction with other words to define the atomic or crystalline building. Example: Microstructure, crystalline structure, atomic structure.

Sea water magnesia Magnesia obtained by a chemical process, essentially using sea water and/or solutions containing magnesium salts, reacting them with limestone or dolomite.

Standardization Branch of human activities aimed at guiding the production of goods and services, in order to achieve: overall economy in terms of human effort, raw materials, energy etc.; protection of the legitimate interests of consumers, through adequate quality of goods and services; safety of people and goods; uniformity of means of expression and communication among interested parties.

Standardization Systematic activity of establishing (specifying) or using standards.

Sintered magnesia Material obtained by calcination of magnesia, or other substances that can be transformed into magnesium oxide, followed by sintering, aiming at densification and hydration stability.

Sampling inspection Inspection of a limited number of items from a lot. See sampling.

Sand A detrital rock, usually siliceous, that after beneficiation is used as an input or refractory raw material.

Sandstone A sedimentary rock made up of grains of silica bound by a cement of variable nature (siliceous, limestone or clayey). Siliceous and clayey sandstones are used in the refractory industry.

Sponge iron Product resulting from direct reduction of iron ore or iron ore pellets.

Simple sampling Type of sampling that consists in taking only one sample per lot.

Simple random sampling Type of simple sampling in which the extraction of n items from a lot of N items is done in such a way that all combinations of n items have the same probability of being chosen.

Sieving Process of granulometric classification of materials and/or raw materials through sieves.

Solid solution Solution resulting from the dissolution in the solid state of two or more minerals in each other, in various proportions and within a range of composition.

Sol Colloidal solution consisting of separate solid particles suspended in a liquid.

Sample Variance Measure of the dispersion of a set of sample data. Sum of the squares of the deviations of each individual data relative to the sample mean, divided by the number of data in the sample, minus one.

Tabular alumina High density granular alumina with low porosity, which presents itself in the form of tabular crystals, obtained from calcined alumina through heat treatment at high temperatures, being practically 100% in alpha phase.

Thermogravimetric analysis—TGA A method in which the loss of mass of a substance is measured as a function of temperature, while the substance is subjected to a controlled increase in temperature.

Thermal conductivity Properties of materials to transmit heat through their mass, from a region of higher temperature to a colder one. Thermal conductivity is expressed in the International System of Units (SI) in watt per meter, per Kelvin [W/(m.K)]

Thermal expansion Property of a material to transduce into an increase in its dimensions as a function of the increase in temperature.

Thermal emissivity The ability of a material to radiate heat, expressed as a percentage of the ideal heat radiation of a black body.

Test Technical operation that consists in the determination of one or more characteristics of a given product, process or service, according to a specific procedure.

Thermal gradient Uniform temperature gradient that is verified in a homogeneous and isotropic medium, when the difference of: The temperature between two points located 1 m apart from each other is equal to 1 K, expressed in Kelvin per meter.

Total or 100% inspection Inspection of all product units from the inspection batch.

Total porosity Ratio of the total volume of open and closed pores of the test body to its apparent volume, expressed in percentage.

Talc Hydrated magnesium silicate with empirical formula $3MgO.4SiO_2.H_2O$, with hardness 1 on the Mohs scale.

Traceability Ability to recover the history, application or location of a product or service, through registered identifications (allows re-verification).

Temperature Measure of the intensity of molecular kinetic activity (energy) such that this energy spontaneously flows from any body at high temperature to any other at a lower temperature in contact with it.

Tridymite Crystalline polymorphic variety of silica, stable between 870 and 1470 °C.

Thermocouple Device for measuring temperatures, which consists of two conductors of different metals, soldered together at one of the terminals and connected by the other to a potentiometer or voltmeter, calibrated, to indicate the temperature of the soldered joint.

Test See assay.

Texture Macroscopic relationship existing between the various forms of pores and grains in a body.

Titania Titanium dioxide of formula TiO_2.

Thixotropy Property that certain materials have to become more fluid or less viscous when agitated than when they are at rest.

Tolerance Permissible deviation in a dimension or property of a material in relation to an established standard or an average value.

Topaz Mineral of composition $Al_2SiO_4.(F,OH)_2$.

Variable inspection Method that consists of measuring and recording the results obtained for a given quality characteristic for each item in a batch; it is usually inspection based on physical, chemical, or thermal tests.

Vermiculite Mica mineral that, when heated, loses water of crystallization and expands in lamellar form; it is a raw material used in insulating refractories.

Viscosity Property defined by the resistance of a fluid material (liquid or gas) or semi-fluid to the movement of internal friction, determined by specific instrumental methods. It is a measure of the greater or lesser fluidity of the material, usually expressed in poise or centipoise.

Viscometer Instrument for measuring the viscosity of liquids at a specified temperature and atmospheric conditions.

Vitreous Partially or completely formed phase of glass.

Vitrification Process of converting a substantial part of a refractory into a glassy phase.

Vitrify To make glassy, usually by heating.

Volatile Relative term expressing the tendency of a substance or a component of a material to transform into gas or vapor at room temperature or in some other temperature domain. When applied to refractory materials, volatiles are those components that, during thermal treatment, transform into gas or vapor as a result of decomposition or other chemical change.

Wet chemical analysis A set of procedures that involve only chemical methods for determining chemical contents, such as titration and gravimetry.

Wollastonite Triclinic mineral of composition $CaSiO_3$. Inverts to pseudowollastonite at 1200 °C; melts incongruently at 1544 °C.

Young's modulus See modulus of elasticity.

Zircon See zirconite.

Zirconia Zirconium dioxide with formula ZrO_2.

Zirconite Zirconium silicate with empirical formula $ZrO_2.SiO_2$ (or $ZrSiO_4$).

Zirquite Zirconium titanate with empirical formula $(Ca.Fe)(Zr,Ti,Tn)_2O_5$.

Bibliography

ALBERS, A. P. F. et al. **A simple method of characterizing clay minerals by X-ray diffraction** . São Carlos, SP. http://www.scielo.br/scielo.php?pid=S0366-69132002000100008&script=sci_arttext Accessed on 31/08/2006.

DANTAS, Elizabeth Sonocia Keiko; **Introduction to Atomic Absorption.** IPEN Course, São Paulo, 1995.

GOMES, Celso B. **Instrumental Analytical Techniques Applied to Geology. Pro-Ore.** Publisher Edgard Blucher, São Paulo, 1983.

MALISKA, Ana Maria; **Scanning electron microscopy and microanalysis.** UFSC, Santa Catarina, 2004.

MOTHÉ, Cheila Gonçalvez; AZEVEDO, Aline Damico de. **Thermal Analysis of Materials.** School of Chemistry – UFRJ, iEditora, São Paulo, 2002.

Technological Characterization of Ores, chapter 3, R. Neumann, C. L. Schneider and A. Alcover Neto CT2004-181-00 Technical Communication prepared for the 4th Edition of the Ore Treatment Book, CETEM, MCT, 2004).

C. P. Bergmann, *Techniques for Analysis of Mineral Raw Materials*, Topics in Mining, Metallurgy and Materials Engineering, https://doi.org/10.1007/978-3-031-58985-0

The manufacturer's authorised representative in the EU is Springer
Nature Customer Service Centre GmbH, Europaplatz 3, 69115 Heidelberg,
Germany. If you have any concerns regarding our products, please
contact ProductSafety@springernature.com

Printed and bound by CPI Group (UK) Ltd, Croydon, CR0 4YY
01/10/2026
02225987-0004